MANUFACTURING MY MIRACLE

MANUFACTURING MY MIRACLE

One Woman's Quest to Create her Personalized Gene Therapy

Jill Dopf Viles

BLOOMSBURY ACADEMIC
NEW YORK • LONDON • OXFORD • NEW DELHI • SYDNEY

BLOOMSBURY ACADEMIC
Bloomsbury Publishing Inc, 1385 Broadway, New York, NY 10018, USA
Bloomsbury Publishing Plc, 50 Bedford Square, London, WC1B 3DP, UK
Bloomsbury Publishing Ireland, 29 Earlsfort Terrace, Dublin 2, D02 AY28, Ireland

BLOOMSBURY, BLOOMSBURY ACADEMIC and the Diana logo are trademarks of Bloomsbury Publishing Plc

First published in the United States of America 2025

Copyright © Jill Dopf Viles, 2025

All rights reserved. No part of this publication may be: i) reproduced or transmitted in any form, electronic or mechanical, including photocopying, recording or by means of any information storage or retrieval system without prior permission in writing from the publishers; or ii) used or reproduced in any way for the training, development or operation of artificial intelligence (AI) technologies, including generative AI technologies. The rights holders expressly reserve this publication from the text and data mining exception as per Article 4(3) of the Digital Single Market Directive (EU) 2019/790.

Bloomsbury Publishing Inc does not have any control over, or responsibility for, any third-party websites referred to or in this book. All internet addresses given in this book were correct at the time of going to press. The author and publisher regret any inconvenience caused if addresses have changed or sites have ceased to exist, but can accept no responsibility for any such changes.

Library of Congress Cataloging-in-Publication Data

Names: Klemme, Heiner, editor. | Kuehn, Manfred, editor.
Title: The Bloomsbury dictionary of eighteenth-century German philosophers / edited by Heiner F. Klemme and Manfred Kuehn.
Other titles: Dictionary of eighteenth-century German philosophers
Description: New York : Bloomsbury Publishing Plc, 2016. | Originally published under title: Dictionary of eighteenth-century German philosophers : London : Continuum, 2010. | Includes bibliographical references and index.
Identifiers: LCCN 2015040044 | ISBN 9781474255974 (pb) | ISBN 9781474256001 (epub) | ISBN 9781474255981 (epdf)
Subjects: LCSH: Philosophers—Germany—Dictionaries. | Philosophy, German—18th century—Dictionaries.
Classification: LCC B2615 .D53 2016 | DDC 193—dc23 LC record available at http://lccn.loc.gov/2015040044

ISBN: HB: 978-1-5381-9722-6
eBook: 978-1-5381-9723-3

Typeset by Deanta Global Publishing Services, Chennai, India
Printed and bound in the United States of America

For product safety related questions contact productsafety@bloomsbury.com.

To find out more about our authors and books visit www.bloomsbury.com and sign up for our newsletters.

CONTENTS

Acknowledgements		vii
Introduction: Olympic-Size Dreams		1
1	Appalachian Medicine	11
2	Witches' Fingers Grab My Legs	15
3	Captured	43
4	Loss of Control	71
5	Killer Instinct	77
6	Un-Walking	91
7	Contains Photo: Do Not Bend	105
8	Deliberations	127
9	Responsible Biohacker	145
10	Going to Pakistan	169

11	Breaching the Ivory Tower of Medicine	181
12	Attempting to Scale Life's Greatest Challenge	193
Notes		199
Bibliography		203
Index		207
About the Author		211

ACKNOWLEDGEMENTS

There are so many who have lovingly guided me to make this book a reality. I want to thank my parents who raised me in a world full of words and beautiful ideas. From my mother, fashioning my kindergarten "masterpieces" into a book of stories to the summer evenings in which my dad invented "Nipper and Spike" stories, unforgettable tales of two puppies in love with their new world. My childhood home was filled with love and laughter, despite the physical challenges we faced. My four siblings, Janet, Aaron, Bob, and Betsy, have provided so much joy and unforgettable experiences, such as our boat rides on Saylorville Lake.

I want to thank my best friend, Suzy Robinette, who always picked me as a P.E. class partner even though we were always going to lose. I'm thankful for my friends at Hillis Elementary who gave me a very happy start in life. A special thank you to Mrs. Buckingham who nurtured my early love of reading books and writing stories.

I owe a very personal thank you to the surgeons at the Mayo Clinic who transformed my life—Dr. Bianco, who extended my years of walking from age 7 to age 33, and Dr. Klaussen, who provided life-saving scoliosis surgery at age 15.

I'm thankful for my pastors—Pastor Litzner and Pastor Rehfeldt, who strengthened my mind and spirit when my body was weak.

I love the memories I made at Des Moines' Roosevelt High School.

Thomas Horiagon, M.D., I am forever thankful that you served as my mentor and my first instructor of genetics. Your kindness and patience impacted my life in so many ways,

For my sisters in Pi Beta Phi, at Drake University and Iowa State University. It's been a treasure to be so welcomed and included.

To the librarians at the inter-library loan desk at ISU, thank you for assisting me with my numerous research requests for medical papers. I was told, at the time, I had ordered more articles than any student they'd ever worked with.

I'm forever in debt to the laboratory of Daniela Toniolo of Pavia, Italy, and the special attention she gave to a teen in the USA, certain she carried a mutation for Emery-Dreifuss muscular dystrophy. Thank you for believing in me.

I especially want to thank Dr. Ben Johnson and the Iowa Heart Center who listened to a young woman, barely out of her teens, as she insisted on sharing research information to aid her father. Without your compassion, we might have lost a very great man far too early.

I am so grateful for my internship in Kathy Wilson's lab at Johns Hopkins University. I learned so much from my time in your laboratory. What a fun summer!

To Dr. Abhimanyu Garg, thank you for listening to my story and believing in me. You are such a wise and generous soul, so in touch with the art of Medicine.

David Epstein, thank you for listening to my story, over and over countless times, and believing there may be truth somewhere in my flood of stories, photos and anecdotes. You were the first to give my story wings—no, make that rocket propulsion.

To the lovely and gracious, Priscilla Lopes Schliep, you had every reason to decline my request to meet, but I'm so happy you listened, and we helped to uncover one of the greatest mysteries of muscle biology. Your laughter is infectious! You are my forever friend!

Acknowledgements

To the entire staff at "This American Life," thank you for selecting my story, making sense of it, and sharing it throughout the world. It was such a rush to work with all of you.

Sheryl St. Germain, thank you for being such a memorable major professor as I completed my thesis. I'm forever in love with the way you blend prose and poetry. Your writing haunts and inspires me, even 25 years later.

Thank you to the students and staff of the University of Iowa Summer Writing Festival. It's a treat to attend every year. You are my people!!

I owe a special thank you to Inez Boyken and her devotion to my writing and speaking career.

Thank you, Lori Walrath, Ph.D. and your laboratory, for allowing me to share in your innovative research. It's such a pleasure to visit each summer.

Kathy Stearman, how fortuitous to be seated next to you in a summer writing festival course. We had so much fun introducing ourselves to each other. I'm forever in your debt as you shared my work with your agent, Alice Speilburg.

Alice Speilburg, and the entire Speilburg Literary Agency, thank you for selecting my memoir to become a part of your wonderful portfolio. Alice, I felt at ease from the first time I spoke with you. You are warm, and intelligent, and thoughtful in everything you do, I'm so happy to have you in my life. Thank you for championing my story.

I especially want to thank my editorial team at Bloomsbury. Anna Keyser, Mikayla Lyndsay, and Jacquie Flynn. You've impressed me every step of the way! You've each carried a vision of this memoir, and your enthusiasm is infectious. You came through when I faced unexpected surgery and delivered a spectacular finished product. I'm so thankful for all of you.

Joselin Linder, you are the goddess of all writing coaches. I've never met a writer so giving, so bubbly, and so welcoming to all who take her classes. A special thank you to the dozens of writers I've met in your class. We survived Covid together and gained so much in sharing life stories.

Kirsten Love and Kate Arnold, thank you for taking such a personal interest in my story, and urging me to get out there, especially in the times I doubted myself.

Most importantly, I thank the favorite people in my world: my husband, Jeremy, and my son, Martin. Thank you for putting up with dirty laundry, messy dishes, and other incomplete chores as I nurtured the story in my head. My life is so beautiful with both of you in it. I love you both forever!

INTRODUCTION
Olympic-Size Dreams

There's a photo on my screen saver of a woman I've not yet met. Her back is to me, but her genial expression radiates 360 degrees. She sports a gray and yellow Nike track uniform, crinkled number five tag affixed to her shorts. Her high, animated ponytail bobs as she addresses the media. It's the summer of 2012, and she has lost her bid for the Olympic Games following an uncharacteristic fall. She's all smiles for the camera, though she won't grace the Olympic podium as her toddler sings "O Canada, Baby," the first version of a national anthem sung entirely by babies.

I don't know who took this image or why they chose to photograph her from behind. The shot floods my limbic system with memories of my late father. I recognize symptoms of my dad in the runner's body—mosaicism of body fat, shoulders tipped slightly posteriorly, hollowed lower back. In her image, I find a miracle of God, and in my prayers, I name her *Priscilla Lost Sheep*.

Her real name is *Priscilla Lopes-Schliep*.

This is place where the doctrine of science ebbs away, and a miracle rises. Though I am not a physician, I've earned my stripes as a "citizen scientist" of my own rare genetic disease. Priscilla is 2009's "fastest woman in the world,"[1] yet she shares striking similarities to my family's hereditary disorder, traits identifiable to a perceptive physician. *How can this be?*

Though it's an infinitely improbable circumstance, if I were placed before a screen, game buzzer at the ready, and asked to strike when I spotted a genetic resemblance from thousands of images of famous athletes, I could distinguish Priscilla in a fraction of a second—the logarithm in my mind's eye as precise as stealth spyware locking in on facial recognition in a crowd of thousands. Could I stake my home equity, my son's college savings, or my husband's retirement on my certainty?

You betcha.

I've spent hours looking at Priscilla's track footage, wondering what kind of life she lives and contemplating what she'd think of a woman, two thousand miles away, clicking on Google Images and spotting family resemblance. I wonder what she would think, what any of us would think, if we discovered an opposite of ourselves existed in the world—her rocket fuel to my body's kryptonite. *Would she want to meet me? Would the secret crush her track career? Is my wish to know greater than her desire for privacy?*

In 1979, I was four years old, searching for my dad among men in the Pepsi line at Lake Ahquabi's concession stand. Like a juvenile emperor penguin homing in on a vocal cue in a colony of thousands, imprinting was my guide. Without matching legs to face, I instinctively found my dad and encircled his legs as if he were a fireman pole, dribbling rivulets of ice cream with no respect for personal space. I knew my dad, and my dad knew me. I was attuned to the unique warmth of his muscles beneath his thin skin like the hypnotic mirage cast by simmering barbecue coals. I revived this lost memento of my childhood in this snapshot of Priscilla. If I could touch her, I would know if she was a reincarnation of my dad.

I realize these are the musings of a madwoman as I conjure our future—me, invading Priscilla's Wikipedia thumbnail; our faces pressed cheek to cheek, gracing tabloid covers in Sweden; our story spawning podcasts from the United States to New Zealand; physicians, up and down the East Coast, taking in our story before morning rounds; our symptoms, woven into first-year medical school curriculum; and my warning averting a life-threatening pancreatic ailment in Priscilla.

The timing of my observation should give anyone pause. My dad died just six weeks before I make the connection with the photo of Priscilla. It could be anyone's assumption that I am seeing things that are not there in a delirium of grief. Even I begin to doubt what I am thinking—to see details of my dad's body, a man who died of muscular dystrophy, restored in a world-class athlete? For nearly a year, I kept my secrets to myself.

April 2014

> *I can run in my dreams, but only downhill.*
> *Spying Priscilla, I rise from the highest point in the stadium.*
> *Taking the bleachers by twos, I descend toward the track.*
> *My legs churning with the frenetic energy of a storm cloud.*
> *Runners part as I cross lanes; officials cower in my presence.*
> *Above my head, there is the glint of steel, and in a moment, it is over.*
> *Possessively, I tuck what I've come for in my pack.*

> *I cast off the memory as the mindless wandering of the subconscious.*
> *I can't possibly be this person who can maim and steal.*
> *Fully conscious, I would never hurt another.*
> *In the moments between wakefulness and sleep,*
> *My hands wrestle the tangled sheets,*
> *Moving with an intensity I can't control.*
> *I assure myself it's only a dream, but to my horror,*
> *I realize I am searching for scissors and a lock of hair.*

The gentle hum of the air-conditioning unit and the piercing rays of sunlight through the drapery folds stir me awake. For a moment, I forget my whereabouts and stare quizzically at the geometric print comforter draping my legs and the spool of tangled sheets spilling to the floor. Drenched in sweat, I'm filled with uneasiness like a gambler facing a margin call. It is then that I remember why I've come to Toronto.

Since toddlerhood, I have displayed symptoms of Emery-Dreifuss muscular dystrophy (EDMD), a hereditary disorder characterized by slowly progressive muscle weakness and cardiac abnormalities often requiring a defibrillator, a device implanted in the heart that delivers a shock to restore heart rhythm. In some cases of EDMD, cardiac transplantation is required. Emery-Dreifuss is an orphan disease, affecting approximately just 1 in 100,000 people.

As the orphan label implies, the rarity of my disease makes it an unlikely recipient of research funding. Connecting with Priscilla and sharing our story on a world stage may represent the best and only chance I have to forge a better life for me and for my friends and family with EDMD, and provide hope for parents of newly diagnosed children, whose first question is inevitably is, "Who's working on this?" It is my dream to launch my genetic disorder out of obscurity and charity-driven fundraising and into the private biotech market primed for genetic-based therapy. To do this, I must first convince Priscilla.

My mother sleeps soundly in the bed beside mine. She has bankrolled this trip on faith, secure in the knowledge that I've uncovered an extraordinary connection to one of the strongest athletes in the world and that somehow, just a few minutes of her time will unlock decades of our family's genetic mystery. I slide from my bed onto my mobility scooter, my lifeline for "legs" for the day. She believes what I profess, that this woman, known for her ability to stride over hurdles at world-record pace, has a medical condition in common with me, though we are opposites. Her leg muscles are whips and mine, the rigid consistency of leather straps. We represent perfect balance in the universe—my yin to her yang.

In her DNA, compared to mine,[2] I've banked on a minute change in one region of a very tiny gene—a genetic twist of fate so rare that it ranks with the likelihood of a winning Powerball ticket—that links our lives together. This gene of interest, nuclear lamin, could be represented as a tiny uninhabited island in the South Pacific within an entire Earth-size target for mutation and the sequencing of this small region of code, all that's required to sort out if we lie together under the same grove of coconut trees or perhaps wave to one another across a beach.

I can never fully understand what it must be like for a child and birth mother to meet for the first time, but I experience feelings ever so similar as I scan the lobby of the Westin Harbour Castle in downtown Toronto. I am searching for Priscilla Lopes-Schliep, Canada's answer to the track sensation of Lolo Jones. Ours could be like any meeting of two people linked by genetics and separated by fate, except for the fact that she is a world champion hurdler and I cannot walk.[3]

Armed with Google Images and a penchant for recalling detail, I am to meet with Priscilla and surmise more about her rare genetics than any physician has before. It is to be the most brazen thing I have attempted in my lifetime. Dozens of ways this plan could derail flood my mind.

I had alternative plans, of course. I could chase her down at a track event . . . *too likely to result in a restraining order*; take a van load of moms in pursuit of a Pampers meet and greet . . . *too likely to frighten*; or send a letter to her track coach . . . *too stalkerish*. After a year of emails and phone calls, David Epstein, sportswriter and author, facilitated my initial contact with Priscilla. In the Hail Mary of social media punts, he condensed my bizarre theory—that our vastly different bodies share a common genetic link—into a 140-character tweet. Fortunately, her agent, Kris Mychasiw, was a prime receiver and facilitated our first phone conversation.

In a matter of moments, I will know if I have made a colossal error in judgment or have uncovered one of the most intriguing discoveries in the field of muscle biology. Like an inmate taking a seat before a telephone cubicle, I'll gaze upon the empty chair beyond the partition, willing it to fill with a polar image of myself, one that moves free from bondage. One thing is sure—I'll be the first to press my palm to the glass. There's always an unbalance of wills in a situation like this. As the second hand of the wall clock passes our meeting time, my soggy Rice Krispies go down like shards of glass. I consider retreating to my hotel room.

"She'll come," my mother offers reassuringly. "Really, she will."

"I only get one chance," I insist. "If I do this wrong, I ruin it for everyone."

I've traveled two thousand miles to unravel decades of genetic mystery, and I believe Priscilla's body holds the most important clue. If what I suspect is true, Priscilla and I entered the world with the same very rare genetic disorder, but due to the difference in heritage, the effects on our muscles took us in the most divergent paths imaginable. For generations, her family, like mine, has knocked on the door of medicine and been turned away empty-handed, and now I've come to seek the truth.

As a young child, I was enchanted by stories of intrepid researchers who braved the Amazon Rainforest in search of rare and exotic plants that could cure disease. In my dreams, I would venture with them, waiting in breathless anticipation as a rare bloom was spotted in the foreground of a waterfall. I would pray for the silence of the wind, timid steps, and then the most beautiful sound as the plant was snatched from the ground, named, and raised toward the sky. I realize now, it is Priscilla whom I have been searching for all my life, but she's not a plant I can possessively stuff in my pack. She's an irreplaceable individual, a creation of nature with a face, and a voice, and a soul.

Vibrations of my ringtone spin my cell phone on the small bistro table. Spying Priscilla's number, I hesitate like one contemplating whether to clip the blue wire or the red wire to disable a grenade. But, as I place the phone to my ear, I am instantly filled with relief. Priscilla is the only person I know who can simultaneously giggle and say hello.

"OK, I promise, we're on our way," she explains. "I was changing one little munchkin, and the other, my little Nataliya, got hold of a bottle of lotion. She's rubbed it onto her hair and the bed rails, and in between her toes. . . . But I promise, we're coming."

Minutes later, I am ushered back to reality by the sight of a woman maneuvering a double stroller through the hotel lobby doors. A bellboy swerves his cart to miss the *Sesame Street* rattle cast onto his path while the infectious laughter of the young mother fills the hotel lobby as she scurries to retrieve the rattle and discarded Cheerios. For a few moments, I follow her steps, studying her hypnotically like one told to "pick up your cousin at the airport"—an individual never previously seen but known through subtle mannerisms.

I carry the hopes and dreams of the dozens of people I've encountered with EDMD. In my pocket, I carry a cashier's check, given in faith by family and friends, to secure Priscilla's genetic testing. In this moment, I believe I can change the world with $875.31, so I swallow my fear and speak the most beautiful name:

"Priscilla!"

Though it was a brisk spring day, I had somehow imagined Priscilla arriving decked out in track clothes. I'd pictured the bulging arm muscles captured in race photos—arms so chiseled and defined, they appeared to not rest naturally at her sides. The woman before me is strikingly beautiful and feminine, and her face glows with excitement. I am so happy I have come!

Priscilla takes a seat across from me at the bistro table and pushes her apricot-colored cardigan past her elbows. As she speaks, my eyes are drawn to the subtle twist of the veins in her forearms. To a casual observer, this would be overlooked, but for me, it is a shared symptom.

Our waiter arrives, and he continues with the conversation begun before Priscilla's arrival. We had been talking about famous people he has waited on in the hotel lobby, and he adds Cindy Crawford and the Indian from the Village People to the list. I wait for him to recognize Priscilla, but he hurries away to set out water glasses for a new party.

"If I'm not dressed in track clothes, people usually don't recognize me," says Priscilla as if she can read my mind. Clad in black leggings and boots, a gray sequined tank, and loose-fitting sweater, Priscilla blends in with the vibrant city life of Toronto.

After ordering a veggie pizza, we begin to sift through the curious life events that have brought us together.

"This disease has everything to do with being female," Priscilla explains as she describes the close bond that she shares with several extended family female relatives also affected with partial lipodystrophy. "We look out for one another. When I got to a certain age, I had lots of questions, so they told me, 'Some of the girls in our family don't develop enough fat.'"

Though men can also be affected by partial lipodystrophy, the condition is far less identifiable. Males inheriting the condition are often regarded as having an "athletic physique," and the disorder is not identified until metabolic abnormalities are revealed upon blood testing.

In our lengthy phone conversations before meeting, Priscilla and I had discovered an unmistakable connection between our bodies, a trait so rare, it's only seen in 1 in 10 million women. We're missing fat in curious places on our body—from our arms and legs, for example—and finding a normal amount on other places—face, shoulders, breasts, and hands. Our discovery was the penultimate of secret club handshakes, a finding no ivory tower of research could breach. If Priscilla and I were each provided an outline of the female form and colored pencils—blue for normal fat distribution, red for missing fat—then asked to fill in our blueprint on the page, when we would put our drawings together, we would find we had created identical maps of the fat distribution of our bodies.

Beyond the physical connection, Priscilla and I discover that we have faced the same psychological challenges as well.

She tells me, "It didn't seem to affect my health, so we didn't think too much of it. We didn't have a lot of money as I was growing up, so we went to the doctor only if something was very serious. But being bullied, that was something I had to deal with.

"There was this one guy in junior high—it seems I always had to sit by him—and he always had something cruel to whisper in my ear. Often, he's call me 'veiny.' I just hated sitting by him. I'd just cringe whenever I saw him."

As Priscilla continues, I realize she has surmounted some of the greatest obstacles one can encounter in life. It seems impossible that someone who faced ridicule could years later demonstrate the greatest of courage under intense media pressure.

I tell her about one particularly mean girl from junior high. She'd wrap her fingers about my elbows or ankles and then sneer at that circumference.

"Your legs look like they belong to an old lady," she'd insist. "How do you even stand on legs like those?"

"I would absolutely love to go back in a time machine with you and beat her up," Priscilla says in jest.

A sleeping Jaslene awakens with a cry.

"I think I'd better check on her diaper," suggests Priscilla as she beckons Nataliya to join her.

I hadn't planned mischief, but I am given the perfect opportunity—bathroom breaks for a baby and a two-year-old can take an eternity. *I have plenty of time*, I remind myself as I watch Priscilla and her two daughters, along with my mother, seek out the restroom.

I begin innocently, my fingers stroking the base of the goblet, and I twirl the stem ever so subtly in my direction. In my mind's eye, I imagine turning it over beneath my chair and spilling the water onto the rug. Gauging the dimensions of the goblet, I suspect the empty vessel would slip unnoticed in my purse. Numerous waiters serve the bustling hotel lobby; a replacement goblet is a mere hand gesture away. *No one needs to know.*

In my regular life, I could never even swipe a candy from the Brach's Pick-A-Mix display. Yet now the urge to strike fills me, this reversal of the law of predator and prey. For the first time in my life, I am no longer the underdog. Priscilla is among the most elite of the world's athletes, but cunning trumps everything.

Once I have such a valuable item in my possession, I can't imagine stowing it in checked baggage. I imagine taxiing down the runaway, sliding my fingers inside my purse, as I strum the contents with satisfaction. Safeguarding exactly what I've come for—a narrow film of saliva—is all I need to unlock decades of genetic mystery.

Airport security is another story. I imagine stemware is not on the list of banned items, but, of course, this would not pass unnoticed through the X-ray machine. Undoubtedly, I would be grilled by a TSA agent. Broken glass could be considered a weapon. Why am I storing this in my purse? Why not in checked baggage? The truth can't help me at this point—it's not every day that you smuggle the DNA of Canada's most celebrated track Olympian through an airport checkpoint. In the absence of an explanation, my goblet would find its way to the "rejected for flight" bin. There, among the ounce bottles of mouthwash and cigarette lighters, I would have to leave what I've searched twenty years to find.

I wonder if someone else in my position could leave something this valuable on the table. Could a scientist who has discovered the remains of the *Titanic* retreat to the surface within arm's reach of the doomed vessel? Wouldn't a person finding a suitcase of money be tempted to possess it "just for a while"—long enough to feel thousands of bills spill between their fingers? This could be my safeguard. I have promised myself I will accept no for an answer. I have even practiced how I will answer if this is the case. Suppose I could tuck this away, perhaps for a day when the not-knowing could push me in the direction of madness. I could find the answer and keep it to myself.

My departure from reality lasted a mere thirteen seconds. Sanity reenters the picture as I hear Priscilla's laughter reverberate into the hotel lobby. Abruptly, I push the goblet away, sloshing water on the white linen tablecloth, and she joins me at the table. A single thumbprint glints in the sunlight, and as Priscilla pauses to take a drink, my indiscretion is wiped away.

1

APPALACHIAN MEDICINE

Before taking off on a thousand-mile round trip journey into the Appalachian countryside, Dr. Alan Emery maneuvered his medical equipment to fit like a jigsaw puzzle inside his vehicle. In the haste of packing, he barely left room for himself and his personal belongings. He carried the expected tools of the practice[1]—a stethoscope, tendon hammer, and ophthalmoscope—but, additionally, he traveled with a laboratory on wheels—spectrophotometer for enzyme detection, centrifuge for spinning blood samples, portable EKG machine to assess cardiac function, and color palettes for the detection of color blindness. He slid into the front seat, hunched over the steering wheel. Despite the cramped quarters, he described his quest as the "Elysian days of medicine," a time when a physician could simultaneously navigate the roles of the trade—laboratory tech, nurse, and genetic counselor.

Guiding his car amid the slick autumn leaves of the hilly terrain inspired trepidation; the clink of the shifting glassware prompted him to grasp the steering wheel tightly. Often, the path was obscure—mere roadways carved into the landscape rather than the paved thoroughfares of Baltimore, with which Dr. Emery was well acquainted. At times, he feared veering into a field, or worse, careening into a ravine. It took him nearly a day and a half to reach the remote multigenerational hamlet plagued for generations by a mysterious muscle malady.

As Dr. Emery pulled alongside the schoolhouse, their prearranged place of meeting, a gentleman appeared, walking with the characteristic waddling gait, his stomach arched before him like the lord of the manor after a satisfying meal: a condition aptly named lordosis. As he ambled toward him, the doctor noted the permanent flexion of each arm, poised as a cowboy draws a gun from a holster—"the cowboy gait," he jotted down in his clinical notes. Perhaps, most intriguing, was the age of the schoolteacher—his fifties—at a time when physicians were telling parents that children diagnosed with muscular dystrophy would likely die in their teens. Yet here was a man who had been graced with the gift of time—time to wed, to pursue a career, to build a family. For Dr. Emery, it was a most remarkable of finds.

After transforming one of the classrooms into a makeshift exam room and laboratory, Dr. Emery enjoyed an evening meal with the schoolteacher and his extended family. Eight men, spanning three generations, joined the revelry of raucous laughter and generous shots of Wild Turkey bourbon as the young doctor gained insight into the disorder that would one day bear his name.[2]

In the early 1960s, the recently established clinical genetics department at Johns Hopkins was experiencing a revolution in medicine. Though more than seven thousand genetic conditions exist in the world, at the time, many had not yet been identified, and little was known about each disease. Rarely, a physician would publish a report on a curious case, and even rarer was the doctor who would peruse the *Readers' Guide to Periodical Literature* and search among the library's reference tiers for an obscure title. Against this backdrop, Johns Hopkins cardiologist Dr. Victor McKusick began the arduous task of organizing genetic disorders by grouping together similar diseases by symptoms—one volume for dominant traits and another volume for recessive conditions, collectively titled *Mendelian Inheritance in Man*, now referred to as "Online Mendelian Inheritance in Man (OMIM)."

Though World War II ended in 1945, the ravages of widespread conflict lingered for years across Europe. Malnutrition persisted in the population as food rationing continued into the 1950s. Bombing raids left thousands without suitable housing, and infectious disease took root in the plight of meager circumstances. Add to this the struggling postwar economy, and the land was fertile with recent medical school graduates eager to pursue clinical studies abroad. Utilizing a research grant

Appalachian Medicine 13

that originally supported chronic disease patients, Dr. McKusick redirected attention to genetic disorders. With the blessing of the National Institutes of Health, Dr. McKusick embarked on a pioneering expedition, recruiting the best and the brightest of Europe in this new era of genetic medicine, among them British neurologist Dr. Alan Emery.

Dr. McKusick emphasized that a physician could not simply set up shop in a genetics clinic and expect patients to come. Immersing oneself in the community, particularly in home visits, was an indispensable tenet of medical practice. The immersion experience could allow a physician to identify other affected family members with very mild symptoms who had previously been considered unaffected. This nuanced pattern of inheritance, even in siblings inheriting the same genetic mutation, was described as variable penetrance.

In these earliest explorations of genetic medicine, the X chromosome was akin to a virgin prairie. A few "settlements" had cropped up—a bleeding disorder, hemophilia; a cluster of genes determining antigens on red blood cells; and a metabolic abnormality, G6PD deficiency. To strike pay dirt was to find large, multigenerational families, seemingly cursed with misfortune to sons, and test for the known X-linked disorders to provide cursory information about where a particular gene might lie along a chromosome. In a 1961 paper titled "A Benign Type of X-Linked Muscular Dystrophy with Unusual Features," authored by Drs. Dreifuss and Hogan, Dr. Emery had hoped he'd uncovered such a windfall.[3]

"The proband [the first patient in a pedigree to elicit the consultation of a geneticist] was a teacher, and I stayed at his house," explained Dr. Emery in a phone conversation with me nearly sixty years later. "He had a great interest in old English folk songs. One of them was about Queen Bess. She was Elizabeth the First in the seventeenth century, and they must have carried on this memory all those years in the family. I was quite shattered. I got to know those people personally, and that was true."

The revelry of the evening provided a prime opportunity for Dr. Emery to examine affected family members. When leaning against a wall or post, arms crossed against the chest, those with the muscle disorder appeared healthy, even vigorous, as rolled shirtsleeves revealed the "Popeye arm deformity"—atrophied biceps like the plumb of a bell in contrast to bulging forearm muscles. However, in motion, the "waddling gait" was clearly present. When seated on the ground, rising from one's feet necessitated crawling to a nearby table or chair for leverage. The youngest boys, ages eleven and thirteen, moved largely unencumbered, except for tightened Achilles tendons that resulted in toe walking.

The young mothers were well versed in the family's medical history: appointments for surgical release of the tendons were already scheduled on the calendar. Dr. Emery couldn't help but notice the ease and candor guiding the family along a path so frequently traveled. The cases were remarkably uniform. This allowed those affected, and the family at large, to adapt to the slowly progressive muscle weakness characteristic of the family's hereditary disease.

Dr. Emery constructed a family tree as he conversed with the family. The colored squares indicated affected males that blossomed with the blessing of children—four in one case, seven in another.

The following two days, Dr. Emery saw patients at the schoolhouse. As he charted results on the classroom chalkboard, he concluded that the symptoms did not appear to be characteristic of any known type of muscular dystrophy. The occupations of the affected men—two teachers, a farmer, and a grocer—were incongruent with the most common and, tragically, most severe type of muscular dystrophy, Duchenne muscular dystrophy.

Upon his return to Johns Hopkins, the presentation of this mildly affected family bordered on the mythological. There was a smattering of applause, a few polite questions, and dubious faces. What could a junior doctor, one currently sidetracked in pursuit of a PhD, have possibly uncovered? Eventually, even Dr. Emery began to doubt his findings. It would be another twenty years before the account was resurrected, and nearly thirty years before cases of female Emery-Dreifuss patients, like me, were reported.

2

WITCHES' FINGERS GRAB MY LEGS

I was named for a girl I'll never know. According to my mother, she materialized like a sprite—a burst of giggles and sand kicked on a beach blanket—as she perused a book of baby names. The stunning teenager represented all my mother wished for me, happy laughter, playfulness, and popularity, a life free from worry and pain.

Perhaps if she'd run a different course, I might have been named something else. I could have been Kayla, or Wendy, or Pamela, or whatever namesake may have come streaking past. It wasn't the girl who drew her attention; rather, it was a chorus of "Hey, Jill! Wait up!" from the boys chasing after her, words that seemed so removed from the nagging fear she tried to silence as she stroked her growing belly and pondered the genetic traits to beset her firstborn.

Years later, when asked what I could have been named had I been a boy, my mother would stumble for words like someone asked to reveal a forgotten dream. I realized there had been no other vision for my reality. Her blessing for me was attached to this beach and the long limber legs disappearing beneath the waves. As the years passed, I, too, would long for a chance meeting with this mirage of me. Always, a part of my soul attached to her like the kiss of a coin before it's tossed in a wishing well.

Ask any member of my family when my symptoms began, and you'll receive different answers; my parents will say age four; my Aunt Ginger will say age one; my Aunt Ann will say that I was a newborn. Ask me, and I will insist I was nearly three. However, if you ask Aunt Janice, the most perceptive member of the family, she will say she knew something was amiss even before I was conceived.

SPRING 1972

"Bob's grandmother walks like a sleepwalker," whispered Aunt Janice as she gently nudged my mother's arm.

My mother's gaze drifted from her bevy of bridal shower gifts toward the clink of overturned china spilling from the dining room. My dad's grandmother Ethel teetered about the table, clearing leftover napkins and crumbs of cake. Her gait was stiff and robotic, her neck and spine hyperextended, and her head cocked slightly toward the ceiling as if she were sleepwalking. She grasped the chair backs for balance, pausing occasionally to recline her rigid, lanky frame against the buffet.

"I'm telling you, something isn't right," Aunt Janice continued.

My mom may have feigned ambivalence, but the observation colored her thoughts for the remainder of the party. Sure, there was an unusual gait, but perhaps she was witnessing the slowing of a woman in her late seventies. Ethel cared for a three-story home and carried baskets of laundry up and down the stairs from the basement. Certainly, this implied she was healthy.

But my mother's curiosity was piqued when she took her future mother-in-law, Dorothy, shopping for a mother-of-the-groom gown. The store clerk frowned in frustration as she remeasured Dorothy and then headed back out to the showroom floor. Time and time again, she entered the dressing room with poorly fitting gowns—a dress fit snugly at the bustline, while the fabric draped over nonexistent hips. A slim-fitting gown fit her waist-to-hip measurement but could not be zippered closed. Finally, my mother suggested the perfect made-to-order buttercream-colored gown.

When shoe shopping, Dorothy insisted on shoes with a heel, refusing any flats passed in her direction. As she shuffled about the cardboard boxes in nylons, it was apparent that her heels hovered an inch above the ground. Dorothy's Achilles tendons could not fully extend her heels to the floor.

These images flooded my mother's mind as she visited her doctor for a routine appointment weeks before her wedding. *Could these mild traits indicate anything of medical importance? Did they have anything to do with her fiancé's unusual gait?*

"Do you have any concerns?" the doctor asked.

The silence in the room hovered like low-lying fog.

"No, there's nothing," she answered, almost apologetically.

My father lobbied for *Roe v. Wade*, the Supreme Court case that legalized abortion in the United States. On January 22, 1973, he celebrated with others at the Iowa State Capitol Building on the balcony overlooking the state legislature.

"I felt it was a good thing for women," he explained to me when I was of age to understand.

A visit to the Iowa State Fair, several years later, prompted a 180-degree shift in philosophy. He walked in one end of a tent a staunch supporter of women's rights, but upon exiting on the other side, he pledged lifelong support to championing the rights of the unborn. In the intervening fifteen feet, he took in the photos of Lennart Nilsson, a Swedish photographer who captured the development of the fetus in the womb. The photographs were first published in a 1965 issue of *Life* magazine, titled "The Drama of Life Before Birth."[1]

My dad opened a copy of Nilsson's companion book, *A Child Is Born*[2] across my mother's belly, matching my development, week by week, like an architect stretching blueprints across a drafting table. Without the technology of amniocentesis or ultrasound, my mother's widening girth indicated all was copasetic.

Ironically, it would be my growth and development—the very marvel that had captured my dad's heart and imagination—that was going haywire even in these earliest weeks of pregnancy. I wonder how my pending arrival would have been perceived if my parents had witnessed the moment the ultrasound wand stalls, chipper conversation ceases, and the ultrasound tech leaves to summon the doctor. The truth is, it can never really be known whether an anomaly of development would have been detectable while I was in the womb. The technology to do so was not commonplace in 1974.

My dad welcomed my birth with a walk from the delivery room to the nursery. Perhaps a nurse paused momentarily, eyeing my dad's unusual gait—the wide-legged waddle of a cowboy stiffened from a long horseback ride; arms, ever bent, as if poised to draw a gun from a holster. Maybe the nurse silently questioned his ability to carry me safely when he lurched as he raised me to his shoulder. He had an ease and a gentle way with me, and he spoke tenderly of my future as he continued down the hallway.

Later, Dr. Robert Anderson, my pediatrician, visited my mother's maternity room.

"A perfectly healthy baby," he proclaimed. "She just needs a small bit of time in the incubator."

On the happy news, my mother completed the cross-stitch sampler she planned to hang in my nursery—Bambi and Thumper peering into a patch of daisies announcing my birth statistics. She spelled out my weight in cursive—*seven pounds, zero ounces*—perhaps to make me seem stronger and healthier than I really was.

My mother's short-lived peace was broken when my fourteen-year-old cousin Michael came to visit. As he raised my pink blanket, he announced, "Hey! This baby is born with six toes!"

My mother whipped the blanket from me and feverishly counted my toes.

"Made you look," he jested, doubled over with laughter.

I entered my home with the pomp and circumstance of a diplomatic motorcade. My father parked his Chevrolet Cordova the wrong way on our quiet side street and opened the car to meet the steps, allowing for the shortest passage from car to bassinette. As my mother approached the front door, my dad shuffled through the autumn leaves, 8mm camera in hand, zooming in on me, tucked in an armful of billowy white blankets.

Once inside, my mother unwound my blankets with the care of one opening a set of china. She introduced me to my nursery—wallpapered walls of frolicking puppies and kittens and white lace curtains trimmed with narrow strips of yellow velvet. I was introduced to my charmed life—my mother, a stay-at-home parent, and my father, a recent Drake Law School graduate entering private practice.

On my first visit to her home for Christmas, my Aunt Ann swallowed hard to stifle a gasp as my diaper was changed. My mother appeared unknowing as she cleaned my baby bottom with a wipe and cheerfully bantered about my Santa Claus outfit. For my Aunt Ann, my appearance was alarming. She pondered my very slight lower half—narrow legs absent the jolly rolls of a baby weeks old. It was the sight of her son Matt lying next to me that confirmed her suspicions—my legs and behind were dwarfed by his, though he was weeks my junior.

"Something is wrong with the baby," my Aunt Ann confided only to her husband as she waved goodbye and watched my parents' car drive off and disappear behind mounds of Midwestern snow.

She did not speak to me about this for thirty-eight years.

My mom had felt worried for a while. There was a nervous lump in her throat as she joined a Mommy-and-Me playgroup, and her eyes casually caught a glimpse of baby bottom as the mother beside her changed a diaper. Was it her imagination, or was I smaller through my backside than the toddlers my age? It wasn't the kind of thing you could ask the closest of friends; sure, you could talk first teeth, lost sleep, lost sex, maybe even chapped and bleeding nipples, but to compare baby bottoms? This would lead to any number of forgotten playdate calls.

It soon became clear that my small build would affect playdates with friends. One morning in the fall of 1977 should have been like any other, the imprint on my mind consisting of the cotton candy fluff of toddlerhood—the plastic cup–stacking, puddle-jumping, spinning-in-circles zest for life just beginning—but that was before my fall.

As I practiced a game of gymnastics with my friend Jennifer, she ran toward a living room armchair, pushed off the armrest as if it were a pommel horse, completed a twist in the air, and landed effortlessly on two feet. I wanted to impress Jennifer more than anything. As I careened toward the armrest, my arms buckled as I tried to spring from the chair. I twisted awkwardly and landed, shuffling feet in a perpetual spin. The thud of my forehead striking a bookshelf signaled that I had stopped moving. I was so horror-struck with embarrassment that I did not feel pain.

Blood seeped into my hair. As Jennifer shouted for her mother, I grasped her arm and pulled her close to me.

"No tell," I insisted.

I was certain that if Peggy, Jennifer's mother, saw me, she would know I was weird. She would carry me out the door at arm's length and hide me under a pile of leaves. I'd never be able to play with Jennifer again.

Jennifer snapped her hand from me and ran to the top of the basement stairs.

"Blood!" she shouted.

I heard the panic in her mother's voice and her hurried steps up the stairs. As she tended to my wound, she sent Jennifer in search of a towel. Jennifer returned a few seconds later tossing two washcloths on the floor.

"Jennifer, big towels!" Peggy shouted. "Big bath towels!"

Minutes later, my mother arrived to drive me to the doctor's office. I was uncharacteristically quiet on the ride as I squished a bath towel to my forehead and stared blankly out the window. At the doctor's office, I was strapped into a papoose. The stitches sewn to heal the gash in my forehead must have stung, but I only remember thinking that I was different.

From this earliest memory, I was aware, even before my parents were, that my body was unlike the bodies of my friends. I had trouble with actions and movements that drew looks of confusion and concern, but I had no difficulty acquiring fine motor skills. No one understood, not even me.

Around this time, my unusual imaginary friend first made an appearance. Undoubtedly, Murphy was named in tribute to my favorite play place—Murphy Park. As I climbed about the park play equipment, my legs tired, and I sought repose on a wooden play fort tucked in a grove of shade trees. Once I was alone, Murphy would pull up beside me and talk to me. I recall little about him except that he was a grown man who used a manual wheelchair.

"Murphy broke his legs," I explained as I pointed to the invisible entity beside me.

"A person usually breaks just one leg," a well-intentioned relative attempted to explain.

"No! Murphy broke his legs!" I bellowed as I struck my thighs with my palms.

In time, the details of Murphy's predicament came into greater focus. As I reserved a place for Murphy at my tea party, the empty space around him was negotiated to account for his wheelchair. It was larger than one would expect—Murphy's legs stuck straight out in front of him with casts extending to his hips.

My parents waffled between engaging in conversation with Murphy and growing increasingly nervous in his presence. To invite what they feared, the unspoken yet unambiguous reality that my legs were weak, to assume a place at the table was to flirt with reality for the first time. Adding to this macabre aura was the fact that Murphy was an entity seen only in my eyes. *How did Murphy appear to me? Was he in pain? What did we talk about when no one was around?* Instead of fostering open dialogue, Murphy drew me deeper into my own private world.

Though Murphy was a figment of my imagination, my tangible interactions with my eighteen-month-old sister, Janet, could not be as easily dismissed. She rocked on all fours and then charged for my castles of wooden blocks. My kicks and punches were no deterrent. She was clearly the dominant and stronger sibling. As we'd pose for photographs, I'd duck behind her, placing my head atop hers as I smiled through gritted teeth—me, a head taller to designate I was the older sibling—that's how I thought it should be until my dad came around from behind the camera tripod to move me back in place.

For my mother, it was an ordinary scene that prompted her to pause in the bathroom doorway. Bare bunned and giggling, Janet and I took turns placing our hands beneath the faucet of running bathwater and splashing one another. We were dressed in identical undershirts of the same size, though I was three and she was one. Often, we were mistaken for twins—an anomaly that never ceased to fill me with frustration—but on that day our differences were notable. As Janet's palm struck a puddle of water on the tub basin, her frenetic energy radiated beneath her skin, jiggling her full and rounded torso. The same motion in me was devoid of animation—my skin hugged my body snugly as if I were wearing a wet suit. The image lingered in my mother's mind long after we were washed and dried—Janet's bare cheeks the size of two cinnamon buns and mine, the size of twin tea rolls.

It was easier for my mother to imagine something was amiss with Janet. She first signaled her presence in the world as a ribbon of red that spun in my mother's warm bathwater. Soon, it was clear that my mother was spotting during pregnancy.

"Don't count on this one," the doctor sternly cautioned.

My mother imagined the small entity within clinging precariously to her womb. But as the second trimester commenced, the baby kicked forcefully, and estimates of the birth weight soared. She was no longer concerned for her baby's health.

"This one's born with her skis on!" proclaimed the doctor as he raised Janet over the delivery curtain and noted her long feet.

At twenty-one inches and eight and a half pounds, Janet's robust body and earsplitting scream were her first calling cards.

That Janet was healthy and I was not was a concept not lost on me. At the age of three, I was acutely aware that I had lost my birthright. As the oldest by twenty-six months, I was meant to be the leader—greater in strength and size than Janet—though it was clear that she didn't look up to me.

Our sibling rivalry was captured in a black-and-white photo in the summer of 1978. I am seated atop the platform of our backyard swing-set slide alongside Janet. The moment the camera shutter clicked, I was forever preserved in a display of anger toward Janet, my teeth bared in frustration and my arms reaching toward her to push her off the slide.

As Janet grew at an exponential pace, my growth percentile plummeted. Even potty training was plagued with peril. My mother recalled searching for me one afternoon as the gentle plea of "Help me, help me," guided her through the house, soft as a kitten's purr. She found me in the bathroom—face and hands squished together, feet pinned to my ears, and my tiny torso swallowed by the porcelain bowl.

Initially, it had been my small build, rather than overt muscle weakness, that troubled my mother. She assumed that increasing my caloric intake and adding nutritional supplements might enhance my growth, but she grew discouraged as the plot point on my growth chart fell below the tenth percentile for weight, even as my height remained at the fiftieth percentile.

My mother, now caring for three children—me, age four; Janet, age two; and my newborn brother, Aaron—sought answers at the pediatrician's office. Undoubtedly, we were a cacophony of health, and Dr. Anderson's suspicions were confirmed as he opened the treatment room door to find me perched on a corner sink in search of gumdrops in a high cabinet accompanied by the chime of Janet's replacing the metal lid on a jar of tongue depressors. My mother fumbled to remove Aaron's diaper and then presented the three of us in open back gown–style to the doctor. But she was relieved of her worries as the doctor told her I was simply a slender child and a picky eater.

It seems odd that my doctor didn't note an overt medical issue in these early appointments; however, on physical exam, I had achieved major motor skill milestones, such as rolling over, sitting, crawling, and walking on schedule. Frequent falls and requests to be carried happened in much larger environments than the home.

As I walked in the hallway with my preschool class, I fell so abruptly that dust filled my nose and mouth, and my palms smacked the linoleum so hard, the sound echoed off the walls. Each fall was accompanied by a ring of squeaky shoes that stopped abruptly to avoid tripping over me. Next, there was the click-click of the teacher's heels as she circled back and invited me to hold her hand at the front of the line.

Attempts to keep up with my peers exhausted me. A simple walk of a few dozen yards had me laboring for breath. My mouth filled with a bitter taste, and a burning sensation ran down the back of each leg. I gladly accepted the offer to hold my teacher's hand, but when I turned back and spied my friends swinging their arms behind me, the feeling of being singled out felt worse than the quivering of exhausted muscles.

By midsummer of 1979, when I was four years old, my mother was confronted with the painful reality that something was wrong with me. On one occasion, she was standing beside the mother of one of my friends, having a conversation. She asked me to be patient as she finished talking, but moments later, the lady, looking in my direction, gasped. Too tired to even stand, I had curled into a ball on the floor of Kmart, my cheek to the cool linoleum.

On another occasion, walking with my mother, I dawdled around the yellow fire hydrant in front of Jennifer's house as I gazed pensively toward scattered sticks along the sidewalk, not wanting to continue home. To me, they looked like gnarled, upturned hands.

"Witches' fingers," I said as we set off again on the sidewalk.

"There's nothing to be afraid of," my mother insisted as she motioned toward the squat brick home neighboring Jennifer's yard. "She's just a kind old woman who lives alone."

Sharp sticks plucked my ankles as I walked. A whorl of windswept dried leaves spun about my feet, and an imaginary hand grasped my shins and thrust me face-first to the sidewalk.

"Were you daydreaming again?" my mother asked as she plucked dried leaves from my clothes and hair.

"No! Witches' fingers!" I shouted. "Witches' fingers grab my legs!"

I was angered that she assigned blame to an absent mind.

My mother swept me from the ground and held me tightly to her body. As she walked briskly, my body bounced against hers. She turned back only once.

Two hundred yards. Five falls. Something was terribly wrong.

I did not witness the phone call that rocked my mother's world. I heard her brisk closing as she mumbled something about a pot boiling over. As I entered the kitchen, I found her huddled on the floor as she cried, head to her knees. Janet pushed a wooden chair around our galley kitchen as Aaron rocked in his bouncy seat. There was no pot on the stove.

I joined my mother on the floor and rested my head on her shoulder. Years later she would tell me about the call.

"Mrs. Dopf, when your husband came to pick up Jill . . . well, now I understand. . . . They walk alike; surely, you've noticed this. Mrs. Dopf? Mrs. Dopf, are you there?"

Moments later, the hum of the phone receiver echoed through the kitchen.

The unusualness of my physical development could no longer remain a silent worry in my parents' minds as we gathered for a family reunion at a state park. On this day, I learned the most intense fear comes out as anger. My grandpa, or "Papa" as I called him, tossed a softball in my direction. I caught the first few lobs of the ball, but with a wayward toss, I stepped off home plate and crumpled to the ground like a tangled marionette.

"Pay attention!" Papa snarled in an uncharacteristically angry tone.

I bristled at his harsh words. Why couldn't he understand that my body couldn't move the same way as the others' did? I wanted to sit in silence in a dark corner of the shelter house so no one would notice me.

The next pitch was too shallow. I leaned forward on home plate, straining to catch it. But at the last second, I stepped forward, my feet meeting shifting gravel, and I suddenly dropped.

The silence of the people at the picnic was palpable. Meat sizzled on the grill. Great-Aunt Theda, my Papa's sister, paused mid-stir as she prepared the potato salad. All eyes were on me, and no one was hungry anymore.

For a long time, I had considered my physical distress to be a problem on my own, but I realized that there was something curious about my dad as well. He would fall with such force that framed photos tilted. In an adjacent room, my mother, frozen and motionless, listened with the acuity of a hunter until the familiar grunts of exertion revealed my dad was upright again. Life went on normally with the understanding that loud noises needed to be announced before they happened. A laundry basket sent luge style down the stairs could be mistaken for a body in freefall.

Dinner was its own adventure. On a nearly weekly basis, my dad toppled the dinner table. As he leaned his body weight onto the kitchen table to rise, Formica quivered, Jell-O jiggled, and table leg brackets groaned. My mother braced herself like a hockey goalie, but she was no match for the table as it collapsed like a heavy child upending a teeter-totter. We shrieked excitedly as casseroles and cups of milk careened to the floor. Later, we shopped for a new table. The store manager looked on in puzzlement as my dad practiced rising to his feet; after trying this on several models, we finally found one strong enough for the challenge.

I realized the limits of my dad's legs as we stood on a wintry hilltop at Des Moines' Waveland Golf Course. His labored breath pulsed against my windswept hair as he held me tightly. We didn't shoot forward like the sure-footed teens beside us, but rather, we ambled forward, inch by inch, his boots digging into the snow. Moments later, we careened down the hillside as I shrieked with joy. At the bottom of the hill, I cried out for more, but my dad tugged the sled toward our car and suggested it was time to go home for lunch.

Later, as my mom unwrapped my mummified form from layers of icicled winter gear, I questioned my dad's hasty retreat from the hill.

"You must not ask him too many questions about his legs," she told me. "You don't want to hurt his feelings."

A shroud of secrecy covered me from that day forward. On the same day that I learned I was like my dad, I learned that we didn't talk about it, which was very confusing to me. Instead of questioning my mom each time I experienced a new symptom, I kept it to myself.

Sparing my dad from feeling embarrassed became my philosophy as well. I pressed my face against the windowpane as I waited for my dad's car to enter the driveway. I summoned a flock of fairies to encircle his feet over the crunching snow, but occasionally, he fell with the force of a stick-bound scarecrow smacking to the ground. I watched helplessly and resisted the urge to cry out. I knew he wanted to fight his battles solo. He crawled toward something upon which to pull himself up—the hood of his car, the trunk of a tree—anything with enough leverage to raise his lanky six-foot frame upright.

As he left for work each morning, I listened for the shuffle-step, shuffle-step as he crossed the kitchen floor. My mother would straighten his tie midstride and hand off a glass of water as he hurried out the door for work. I slid my feet into a spare pair of his shoes and attempted to make the shuffle-step, shuffle-step just as he had done.

In court, his shuffle-step, shuffle-step echoed off the cavernous marble pillars, and his lurch was melodic and steady like the most reliable of metronomes. Seated in the back row of the courtroom, I watched transfixed, my chin propped on the bench before me, as my crayons and coloring pages slid to the floor.

"You know you're really important if your voice echoes when you're at work," I whispered to Janet.

The concept of disability was lost in the spirited games we played. Upon his return from work each Friday night, my dad would assume the posture of a gorilla. As he ambled about the kitchen, knuckles dragging on the floor, we barricaded ourselves beneath the kitchen table and shrieked with joy and fear.

A scar on my dad's left biceps muscle, which looked like a caterpillar, was the first signal of danger to me. My dad spoke of it infrequently, but he would simply flex and release his muscles, making it appear that the caterpillar was wiggling but traveling no real distance. The doctors had wanted to know more about him, he explained, to satisfy my questions.

As a four-year-old, the true meaning of the scar disappointed me—the caterpillar's legs were not legs but, instead, stitches, and its bulbous body, a testament to what had been taken—a slice of his muscle. On one delightful, rainy afternoon, with magic marker and glitter, I transformed the scar into a butterfly with gossamer wings.

Each Sunday, we occupied the "cry room" of our Lutheran Church. Tucked behind glass, I passed the time pressing keys on the typewriter, and Janet pushed a folding chair around the room. Occasionally, a parishioner gazed back toward the palm-printed window separating the office from the sanctuary and caught a pantomime of the chaos within.

With the offering of Communion, we ventured into the sanctuary, the colors and sounds springing to life. As my dad knelt at the Communion rail, the soles of his dress shoes spoke volumes about his life—the outside rim of each shoe was scuffed and weathered from the supination of his feet and the heels that never touched the ground, smooth as polished stone. A tuft of sock poked through a hole in his left shoe, his dominant leg. He never spoke of why he walked differently, but as he kneeled before God, he was an open book.

1966: DRAKE UNIVERSITY

My future father's fellow cadets in the ROTC didn't ask about the demerits he stored in his locker. He accepted them wordlessly as the captain noticed his heels were not cocked together when he stood in formation. Once, they heard him mention something about a brush with polio.

He wouldn't escape the attention of the visiting physician. As the doctor walked slowly up and down the row, he paused and eyed my dad's arms with curiosity. His diminished biceps dangled like the clapper of a bell in the sleeve of his white T-shirt, though his muscular forearms and hands bore a striking resemblance to Popeye, the Sailor Man.

"I need to see you in my office," admonished the doctor.

It was an appointment that my dad didn't keep.

The next year, he slipped out of the gym door upon the physician's return. In brisk, stocking footsteps, with the locker room in sight, he collided with the doctor as he turned a corner.

"I remember you," the doctor insisted as he studied my dad's arms and noted they didn't straighten.

This time there was no escape. In a field in which fallen arches or anything less than 20/20 vision is cause for dismissal, the doctor's report was career ending—my dad could continue as a flight mechanic, but he would never be accepted into the Air Force as a pilot. Instead, he declared a political science major and planned for a career in law.

Years later, my father told me about the first time he saw my mother, her long blonde hair held back in barrettes to reveal her face, as he spied her from the library reference tiers. He'd pondered the perfect excuse to speak to her before he arrived at the library counter with reference books selected for photocopying. She must have wondered at the curious plethora of academic articles laid out before her . . . legal jargon, mating rituals of dragonflies, import/export reports of Colombian coffee beans. Perhaps she raised a questioning eyebrow as she pondered the volumes that he had so quickly selected from the shelves. This was of no consequence to my dad. He watched with breathless abandonment as her long hair spilled over her shoulder and across her back as she completed his copy requests. As the final minutes of the 1960s ticked past, they set their first date in the new year.

Initially, my mother noticed my dad's hip bumping hers with each step as they crossed the Drake University campus, but she considered it a minor nuisance. She considered asking my dad about it but found herself at a loss about what to say. As a tall and dashing political science major, his waddling gait seemed out of step for his commanding demeanor. He had a way of dismissing questions even before a word was uttered.

She didn't pry into his hasty retreats from games of touch football. She didn't question why he turned backward to walk up steep hills. Instead, she focused on his crystal blue eyes and his lofty dreams for the future.

WINTER 1979

In Sunday school, I sang "This Little Light of Mine" with my class of four-year-olds. My favorite part was the jubilant "No!" as we were asked if we should "hide it under a bushel." With each refrain, I jumped sideways, bringing me closer and closer to the bulletin board. In the chaos of coat collection time, I tugged my coloring sheet from the wall, folded it, and hid it in my mitten. My Sunday school teacher eyed me suspiciously as I backed out of the room, hands behind my back. Safe inside our station wagon, I drew the cuff of my mitten over my nose and mouth and inhaled the aroma of damp manila paper and melted crayon wax. *Perhaps it was already working,* I tried to convince myself.

As we returned home, my dad played 8 tracks of gospel music, the heartfelt pulse vibrating the windowpanes of our living room. He opened the windows and doors and released the strains of "How Great Thou Art" into the yard.

"Bob, the neighbors!" my mom said.

"But that's the point!"

Our favorite was Frankie Valli and the Four Seasons. He would crank up the treble falsetto of "Rag Doll," and we would twirl about the room.

"I have Jesus in my mitten," I explained to my mother as we ate our traditional pot roast dinner. It took some convincing, but I was finally allowed to wear a single mitten at the table.

"You mean you have Jesus in your heart," she countered.

"Nope, he's in my mitten," I confessed with gusto.

After nighttime prayers, I pulled my coloring sheet from my mitten. It pictured a man in a purple robe who leaped in the air, his cane tossed to his side. A throng of jubilant followers extended their arms to the sky.

"Jesus heals the faithful," the caption explained.

I wondered if Jesus was like Santa Claus. Maybe he needed a wish list, maybe cookies and a glass of milk, or perhaps he just needed to know you were sleeping before he visited. I circled the healed man with a lasso of crayon and fell asleep with the drawing on my chest.

As I woke in the morning, I lay perfectly still, wondering if I could possibly sense energy pulsing through my limbs. I started with my toes and wriggled them with renewed frenzy. Then I imagined my body was suddenly jolted with amazing strength. But as I leapt from my bed to show my family that I was healed, my words were cut short as my stockinged feet met the hardwood floors and I slipped and fell.

The following night, I appeared like a whisper in the kitchen doorway.

"Will my baby be like me?" I asked my mother.

"I would certainly hope so," she said after a brief pause. "You're a very bright girl. All your teachers say so."

"No!" I shouted, a bit too loudly. "Will my baby have legs like me?"

The question had come far too soon. My life had been peppered with too many medical terms. I knew I was like my dad, and naturally, I had begun to wonder about the future.

"It's like this coin," my mother said as she picked up a coin from the kitchen counter. "If you flipped the coin, you could get heads or tails. Each one is as likely as the other. This is how it would be if you had a baby. The baby has an equal chance of having weak muscles or not having weak muscles."

I frowned as she placed the coin in my palm.

"So what about Janet? Could her baby be born like me?"

"No, not for Janet."

This time the words spilled quickly.

"It's not fair," I said. "I get all the hard things and she doesn't get any."

My mom took me back to my bedroom and tucked me into bed.

"Is there anything else you wanted to talk about?" she asked.

"No," I lied.

After the door closed, I slipped out of bed and studied Janet as she slept. We were dressed in identical Mickey Mouse nightgowns, though mine was pink and hers was yellow. Her shapely arms were hugged by her elastic cuff, while mine swam in my cuff like a broom handle.

Settling back into bed, I watched the headlight of passing cars circle my bedroom walls. With each burst of light, the painting above my bed sprang to life—Jesus, the good shepherd, surrounded by a flock of sheep. I kneeled on my pillow, placing my hands on the wall on either side of the blue frame.

"How could you do this to me?" I asked. "I just *got* here."

My mother had avoided saying the words for as long as she could. Nothing put her stomach in knots more than the annual telethon for the Muscular Dystrophy Association[3]—the slapstick comedy of Jerry Lewis, firemen pledging to "Fill the Boot," celebrities interrupting the somber undertones with jokes, adorable children who walked like me crossing a Vegas-style stage, sobbing parents interviewed by celebrities, and soda pop bottlers entering the melee with six-figure checks to help "Jerry's Kids." It was a jarring, heart-stabbing, mind-numbing extravaganza that arrived, without fail, every Labor Day weekend.

"My kids can't walk . . . my kids can't take care of themselves . . . my kids can't work," pleaded a perspiring Jerry Lewis as he labored through his twenty-hour telethon and pointed to his tote board, his beacon of hope.

"Just one more dollar than last year," he urged. "That's all I ask for my kids."

It was no wonder both of my parents shied away from the premise of the telethon. As a federal prosecutor, the plea to "save Jerry's Kids" diminished my dad's profession. The dour mood of the fundraising was in stark contrast to my family's upward trajectory in the face of life's challenges. Sadly, the emotionally overinflated telethon was the only exposure my mother had to others with my curious gait and was fodder for the questions she posed at the pediatrician's office.

"Could she have muscular dystrophy?" my mother asked the nurse at a routine appointment when I was four years old.

"This one?" the nurse asked with a puzzled face. "No, it's only the boys that get that."

I imagine my mother enjoyed a moment of reprieve. The nurse was unambiguous in her response. In this small exam room, I "passed" as a "normal child." My mother may have chastised herself in the intervening minutes—she was tired and overly worried. There was a simple explanation—maybe I just needed more time to sort out my gross motor development. She would ask the doctor for some exercises.

As Dr. Anderson entered the room, it was clear by his somber expression that this would be a difficult appointment. Undoubtedly, he had had concerns about my physical development for some time, but as my mother would soon learn, there were many aspects of my case that were atypical. The first line of the Hippocratic Oath, *First do no harm*, likely served as guidance; he didn't want to make such a severe diagnosis unless he was certain. Perhaps he had been waiting all this time for my mother to bring up the difficult questions.

I perched unknowingly on the exam table swinging my skinned knees as I sorted a bottle of gumdrops, setting aside all the red ones.

It would seem logical that there would have been a high-tech battery of tests to check for muscle dysfunction—maybe a scan of my body from head to toe, plotting a three-dimensional view of my musculature, or a machine to test how many newtons of force I generated with a push or kick. But instead, Dr. Anderson made a simple request of me: "Be a bird."

I looked on in bewilderment as Dr. Anderson took a seat on the exam floor and invited me to sit beside him. He extended both arms laterally, like a bird in flight, and rose effortlessly to a stand.

Little did I understand that my movements of the next few moments would predetermine the course of my life. I placed one palm atop the other on bended knee and then pushed downward with both palms as I leveraged myself upright.

"Let's try again," prompted Dr. Anderson as he ushered me to the floor once again. This time I was prompted to begin with my arms in full extension.

"There you are—be a bird, be a bird . . . don't let your wings fall," prodded my doctor.

I looked on in confusion. My bent knee quivered.

"That's my handle," I explained as I pushed off my knee again.

Dr. Anderson explained to my mother that I had a positive "Gower's sign,"[4] which is a type of bracing displayed by a patient with a muscle disease when asked to rise from the floor. First described in 1879 by a French neurologist, the physical test had become the gold standard in identifying cases of muscular dystrophy. The test is especially valuable in the case of pediatric patients because young children typically "spring" from the ground without impediment. The "pooled in molasses-like" rise, typical of neuromuscular disease, is disheartening, even to a novice observer witnessing an affected child among a group of "normal playmates."

Additionally, my blood draw revealed an elevation of an enzyme called creatine kinase, or CK. An elevated CK reading is typically associated with active muscle disease; however, it may be raised following aggressive injury to muscle as in the case of a boxing competitor following a match. An elevation of CK may be observed following a heart attack as damaged cardiac tissue leaks the enzyme into the bloodstream. My CK was elevated to 818 U/L (units per liter), which was significantly higher than a normal reading of less than 200 U/L. Dr. Anderson explained that he followed patients with Duchenne muscular dystrophy, a much more severe type of muscular dystrophy, featuring CK readings between 5,000 and 150,000 U/L.[5]

Rather than being a single disease entity, the term *muscular dystrophy* encompasses nine main types of muscular dystrophy. Some forms affect the heart as well as voluntary muscles, and the condition may or may not be life-threatening. All muscular dystrophies have a genetic component. Some cases are due to a faulty gene in two "normal" parents, though in some cases, having just one misspelling in a muscle gene could be enough to pass on a muscle disease.

As I was examined on the table, Dr. Anderson explained that several of my symptoms were familiar to him—the Gower's sign and the waddling gait of my walk; he had seen these peculiarities in his patients with Duchenne muscular dystrophy, a very severe form of the disease.

"This is not typical," he explained as he pointed out the characteristics of my leg muscles, drawing them out in extension and bending the knee of each leg in flexion to my chest as I laid on the exam table.

"In Duchenne patients, I see large, bulky, lazy muscles; these are small and tight."

He noted my chart and then spoke directly to my mother.

"This is far beyond my expertise," he explained. "I'm going to need to send you to the Mayo Clinic."

"Can I have all the red ones?" I interrupted as I pointed to my pyramid of red gumdrops.

"Yes, you can have all the red ones."

Dr. Anderson closed the appointment in the same manner he would for every appointment of mine until I was nearly nineteen. Over the years, my medical file would bulge to the point of nearly overflowing, stuffed full of documentation of how I was different. With each closing, he would tap the folder lightly on the top of my head, making sure to make eye contact with me.

"You're a beautiful girl," he'd say.

"And now, Miss Jill, shall we dance?" questioned Dr. Hymie Gordon, chair of the genetics department at the Mayo Clinic.

Dr. Gordon bowed to me and lifted me onto the exam table. Standing on my tiptoes, my eyes aligned with his. Gently, he guided my movements like the rise and fall of a carousel horse.

"You see, I can get a sense of what I need to observe, and we can make a game out of this," he suggested in a singsong tone.

His raised arms guided my pirouette as he winked at my parents.

"Just like this, I can gauge her strength and flexibility."

I imagined I was a miniature ballerina in a jewelry box. He directed me to raise my arms high over my head, and I squealed each time I spied my image in the mirror.

"Note, lordotic posture is present," he indicated to the charge nurse as he pointed out the inward curvature of my lower spine and the tendency for my backside to stick out.

"Very good. Very good," he said as he directed me to take a seat.

As the doctor returned to his stool, I climbed onto his lap and seized the chance to run a toy car across his shoulders. My mother attempted to coax me to her side, but Dr. Gordon refused the offer.

"No worries," he insisted. "All of this is important for observation. You see, she climbs about just like any young girl her age."

My mother's face softened like one receiving a compliment.

"However, Jill has a positive Gower's sign," Dr. Gordon continued. "It's a type of bracing a child with a muscle disease demonstrates when asked to rise from the ground."

I mussed the doctor's hair as my two-year-old sister, Janet, accidentally smacked her head with a toy train and wailed at the top of her lungs. My newborn brother, Aaron, awakened in his baby seat, and I was returned to my mother's arms.

"It's OK. No worries," insisted Dr. Gordon. "This is the world-famous Mayo Clinic. We can solve any problem."

He slid across the exam room floor on his low stool with wheels.

"I need lots of friendly, smiling nurses," he said as he pushed his intercom button, "and lots and lots of toys."

Momentarily, the door opened to reveal two young nurses and a basket of toys. They took a seat on the floor and engaged Janet with a spinning top.

Dr. Gordon explained that my father's case was not caused by polio. An enzyme called creatine kinase, or CK, indicated an active muscle disease was present. Normally, CK is found within muscle cells, but upon disease or injury, the enzyme may leak into the blood. The CK elevation indicated a diagnosis of muscular dystrophy, though the exact type was not known.

Our extended family's medical history was explored next. Dr. Gordon noted that my father's mother and grandmother developed type 2 diabetes in their fifties and my mother's father was described as a "brittle diabetic" and was insulin dependent from the age of thirty-five.

"My mother says she has legs that are not full," suggested my dad.

"Could you expand on that?" queried Dr. Gordon.

"She's always been proud of her legs, as she has very little fat; they look muscular."

As I will learn in future classes in genetics, a thorough genetic evaluation involves exploration of extended family as well. Dr. Gordon inquired if these women may be available for examination.

"No, this has nothing to do with them," my dad insisted as he shook his head. "My grandmother passed away a year ago. They don't have any muscle weakness whatsoever."

In my paternal grandmother's living years, I would share my dad's opinion. She showed no sign of weakness, and in photos I had seen of her early years, she blended in with other young women her age in tennis matches and dance lines. Still, I have the curious memory of her as she climbed the ladder of my parents' backyard pool and seeing the backs of her thighs, which were smooth as polished stone with no hint of cellulite.

Dr. Hymie Gordon, Mayo Clinic geneticist. Clinical notes: 8-20-79

> "In the family history it is recorded that the paternal grandmother [II.4] Dorothy Dopf is said to have 'leg muscles that are not full.' Moreover, the paternal-great-grandmother [I.2] Ethel Desmond is said to have had 'flat-feet' and to have 'walked with a sway.' It would not surprise me if these two relatives had this same kind of proximal muscular dystrophy with very mild expression."

Though I do not have a memory of this, numerous measurements were taken of my anatomy. Hereditary disorders can be visible in the face, and deviations from expected proportions of limb segments can occur. It seemed odd and a bit macabre to search for anatomical differences in a new life just beginning, and I imagine my parents may have been unnerved by the experience. However, they sought counsel like refugees from a war-torn nation finding refuge in an embassy. They assumed all this cataloging and dictation would bring us closer to the all-important phone call as our family's medical mystery was solved and we were advised of treatment options.

"But what about the children? What about when it comes time to start a family?" my mother inquired.

"Mrs. Dopf, this is needless worry," said Dr. Gordon assuredly.

I reached for his stethoscope, playfully batting it back and forth across his chest.

"The answers will come in time. Someone's going to solve this mysterious disease."

Poolside at the Holiday Inn, my mother listened to the other mothers talk about the maladies that had led them to the Mayo Clinic. She uttered the words "muscular dystrophy" for the first time aloud, and her eyes met a trio of horrified faces.

"Your husband . . . and the baby?" said one woman as she shook her head in disbelief. "Like those poor, poor people on the telethon?"

My mother tried her best to redirect the conversation, focusing instead on my father's burgeoning legal career.

"Whatever are you going to do?" questioned another woman as she leaned in closely, clasping my mother's hand.

I shrieked with joy as I tipped a bucket of water on my dad's head and then danced a brief jig, slipping in a puddle of water along the pool's perimeter. My wails filled the chlorinated air and reverberated off the shallow ceiling.

"Not the little one in pigtails too?" one of the women gasped.

No amount of redirecting could save my mother now. She'd been deemed the one most worthy of pity. Later, as she packed away our swimming gear, she looked back at the empty lounge chairs with chagrin, thinking *never again . . . never again*. At that moment, she vowed to keep our family's medical problems a guarded secret.

The pediatric neurology department at the Mayo Clinic was far too still and quiet. Children's books were stacked neatly on a small shelf and the toys and cushions, too organized and clean as if no child had leapt about them. The nurse, clad in a white hat and cape, typed clinic notes as she gazed toward the sprinkling of waiting room patients. I was bored and listless, and my only source of entertainment was my squeak of protest as I slithered about the faux leather seat.

To ease my boredom, my dad beckoned me to a mysterious piece of wall art. The step stool before it raised me to eye level. Side by side, our hands explored the tempered glass, steering the purple goop inside into an ever-changing impression of ourselves. I was too young to understand the inner workings of the art—a metamorphic sculpture stirred to life by the heat of our bodies. What I did know was that a few well-placed palm prints could transform into the flight of a dove or reveal the hidden turmoil within storm clouds. My face and that of my dad could spin about one another like the rolling cadence of surf. In this world, I could control motion with a simple slide of my hand. If something displeased me, I could simply wave it away.

Moments later, my name was called. For a few brief seconds, I harnessed all my energy into the frenzied stirring of hands, elbows, and face. As I approached the nurse's station, I turned back to gaze upon my work, satisfied to see my creation churning even as I was about to disappear from the waiting room.

"Wait!" I suddenly called out.

The front desk clickety-clack stopped, and several patients stared in my direction. The nurse glanced at her wristwatch.

"I want to see what I become!" I announced.

In a moment of serendipitous love, my dad gently raised his palm to the nurse and turned to see our picture. We were holding up the progress of the most famous clinic in the world, but we turned our backs to watch our art stall like cooling lava.

I had not yet learned to despise my visits to the neurology department. I savored the melodious click of my shoes on the parquet floor and the sight of a tricycle and a red rubber ball at the end of the hall.

"They color pictures of me!" I excitedly whispered to my mom as I passed by on the trike.

I loved the attention. An intern in a long white coat accompanied me as I played, his exuberant pen scribbles testament to the art I inspired. Next, I was asked to climb a set of stairs to nowhere. As I paused on the landing, I saw the picture the doctor was decorating—an outline of the body covered with bold pen markings and long words that spilled into the margins.

Later, I was invited onto the lap of the neurologist for a story. Naked, except for my exam gown, I did as I was told. Minutes into the story, I felt a strange sensation in my bottom. I looked in horror at the doctor and then searched my mom's eyes for answers.

"She won't remember any of this," the doctor explained to my mother. I'd been given an anal suppository.

The room began to swim, and I returned to my mother's lap. The doctor exited the room, and a nurse, wearing a freshly starched white cap, entered. She prodded me to accept a spoonful of pink medicine, but my head hurt and I felt very cross. So I batted it away with my fist. She poured another spoonful and pinched my mouth open with her thumb and index finger.

"There. That wasn't so bad," she said. Her fingers pressed into my face, and she smiled with satisfaction.

For a split second, she sensed the sucking of my cheeks, but it was too late. Frothy pink liquid spewed from my lips, coating her face and hair. The smile of satisfaction was all mine.

In future years, I'd look back on the timing of this fit and realize it was one of the most fortuitous of my young life. I had missed an appointment with an EMG machine, a medical device that utilizes needles placed into muscle and stimulated with electricity. The test is used to determine if a disease is myopathic or neuropathic. In an adjoining room, my dad, though sedated, had gained sufficient alertness to realize the horror of this painful test and had begun searching for me.

"You're not doing this to her!" he called out as he reached the hallway.

His words were slurred, but I recognized my dad's voice. Seconds later, he appeared in the doorway. Wires and suction cups sprang from his gown. An orderly chased after him, calling out for him to sit—his sedatives had not yet worn off. He surveyed my exam room quickly, noting an overturned chair and an angry nurse with a lopsided nurse's cap sprayed in pink.

"Good girl!" he exclaimed.

We returned to the hotel and slept for hours. Later in the evening, my mother attempted to take us to dinner in the hotel restaurant.

"I want a highchair," I drawled to the woman at the hostess stand.
"But, you're four, honey," my mom explained.
"I want a high chair!" I wailed.
Guests in the dining room peered in the direction of the lobby.
"Right this way," the hostess said, retrieving a high chair on the way to our table.

Though we were placed in a darkened corner of the restaurant, my mother's chagrin was palpable. She prodded my dad with her elbow as he began to fall asleep and then gave up as he rested his head on the table beside her soup. I stirred saltine crackers on my high chair tray, occasionally crying out in unintelligible moans.

Looking back on this experience with adult eyes, I imagined my mother, barely twenty-nine years old, may have considered this an omen for her future. My dad and I were tethered to her, unable to handle the most mundane tasks independently. In a matter of hours, my dad had been transformed into someone she couldn't converse with, and my crying out reminded her of her previous work as a special education teacher. I suspect she would have reminded herself that our afflictions were only temporary. My dad would resurrect into an attorney with a magnetic presence, and I would challenge my mother to a game of Memory.

The neurologist was taking too long with my sister Janet's exam. With her head tilted to the side in deep reflection, the neurologist palpitated below Janet's jawbone. She excused herself and returned with two additional physicians.

"I've never felt glands like this in a two-year-old," remarked one of the doctors.

My mother was handed a new packet of appointment cards—this time for Janet. Unlike the cards for my dad and brother and me, these were stamped with a single word, *RUSH*, and we ventured from the neurology department to the oncology department.

I can never begin to imagine the horror my mother must have faced as she was led into yet another waiting room, this time filled with young children with sallow faces and bald heads. I stayed in the waiting room with my dad and baby brother until Janet and my mom emerged well past closing time—my mother rubbing her temples and my sister's arm lined with Band-Aids of cartoon heroes. Janet had received a clean bill of health, but my mother's emotional state had taken a serious toll.

My parents checked back into the Holiday Inn and unloaded our packed belongings. My mother showered and fell into bed beside me, a lump of terry cloth robe, violently shaking. My dad dimmed the lights, and we watched *Little House on the Prairie*, on mute. The dim stream of light beneath our room curtains was the only connection with the outside world. Intermittently, plodding feet ran past our poolside room, inflatable toys squeaking as they brushed past.

3

CAPTURED

If I could travel back in time and relive a favorite childhood memory, I would return to a summer day in 1980 when my parents washed our station wagon in the driveway and soapy suds spilled into the tire tracks in the grass. My brother, Aaron, and I pushed our Matchbox cars through the muck, our bodies caked with mud as if we'd acquired a second set of clothes. On this day, our muscles were as pliable as the mud that squished between our toes. We had yet to learn that in future years, our once supple muscles would become as taut as leather straps.

When I was old enough to understand, my mother explained that like me, my brother was born with weak muscles. He toddled about like any other child except his knees would stick out laterally as he ran. He was stronger than me, the doctor explained. I had never felt the motion of running, so I watched him with much interest.

Over the years, I sought significance for the birth order of Aaron and me—siblings born on the same day, four years apart. I adopted the concept of "Leap Year Twins," as testament to our connection, our paired entrance into the world as precise as the mathematical certainty of the Earth's journey around the sun over four years' time. I imagined we had the same spiritual connection as that between brother and sister pairs of figure skaters, sensing one another's movements and hesitations. It was my wish to be reunited with him someday in heaven, like athletes trained in the most arduous conditions and then returned to sea level, lungs surging with the sharpness of undiluted air.

I was too young to understand the nuance of genetics, but I sensed danger in our matching coloring. We shared the same ivory skin and halo of wispy blonde hair that bleached nearly white in the summer sun. When my second brother was born in 1984, I slipped unnoticed into his nursery and held him in the dark. Softly, my lips nuzzled the fuzz atop his head, a whisper of what lay below the surface.

"Don't have blonde hair," I said. It hadn't escaped me that my and my brother's blonde hair was spun with the tripping and falling spell.

The first thing my mother did when we returned home from the Mayo Clinic was to place me in the hands of strong women. She began with Sue White, the director of Drake University's aquatics program and provider of the most skilled private swimming lessons money could buy. As our station wagon entered the secluded gravel driveway, a kaleidoscope of images, suggestive of a fairy-tale garden, splintered the privacy fence. The drive circled to the back of the property, where a row of wooden benches beneath a grove of trees shaded the watching mothers from the noonday sun. They were greeted by a view of the large oval pool, glittering with sunlight. Above it all, the diving board stood sentinel like a lion at the head of a pride, unfurling its turquoise tongue to cool and to warn.

The surrounding foliage was filled with all manner of butterflies and songbirds, and I eagerly pumped my legs as the creaking of the garden swing announced my presence. I could live to be a hundred and one, and I would never forget the majestic call of my name, my sandaled feet striking the red earthen bricks—clop-hiss, clop-hiss, clop-hiss—that circled the perimeter of the pool and led me to my swimming instructor, Sue White, her skin the color of several swipes of my burnt sienna crayon, her tousled hair tamed by a visor, and the tip of her nose graced by a dollop of sunscreen. She had the air of a woman who did not intend to wet her hair except in the direst of circumstances.

She met me at the water's edge and swiftly placed me on a concrete block in the shallow end of the pool, the tips of my shoulders and pigtails kissed by the lapping of the cool water. We clasped our hands together and counted in unison, "One, two, three." Then I was pulled under the water, my bangs pooling at the surface like a silken lily pad, toes struggling for a foothold on the unforgiving block. As quickly as I had been dunked, my arms were tugged to the surface, and I emerged sputtering. Chlorine burned my nose, and my fiery eyes searched the grove of trees for my mother. *Had she seen this?*

It occurred to me then that I didn't have that kind of mother and never would, the kind that leaps from a picnic table and races across the park, first-aid kit in hand. My tumbles off the end of the slide weren't greeted by rubbing alcohol and a sterile Band-Aid. I was never urged to nap in a quiet shelter house after lunch or accept a head-start push from my mother at the Big Wheel Rodeo.

"You're angry," Sue White prodded.

Wordlessly, I gritted my teeth and nodded in agreement.

"Now, take all that anger, push it through your body . . . down to your toes. Then I want you to blow, blow, blow!"

Momentarily, I was tugged underwater again, but this time, I anticipated the flood of water.

Blow. Blow. Blow. I sent the bubbles tumbling this time.

At the close of our lesson, I asked about the diving board.

"You can take a walkabout," Sue White explained, "but to jump off the board, you are going to have to show me you can swim from one end of the pool to the other."

I did a quickstep around the pool, careful to contact the sun-warmed concrete as briefly as possible, then felt the cool of the diving board beneath my toes. At the edge of the board, I peered into the water, a no-nonsense descent of twelve feet. It will happen someday, I promised myself.

In contrast, the physical therapy sessions prescribed by my neurologist held none of the charm of Sue White's lessons. As I sat on one of the steps, water lapping up to my ankles, I was encouraged to play a ring-toss game, the target floating in the water two feet from me. The rings were squishy and of many colors—the kind of toys babies chew on. Two women joined me in the water, their laughs echoing off the indoor aluminum pool of lukewarm water.

"This is a baby game," I protested. They paid scant attention to me and continued their conversation.

After I was dried and dressed in my workout clothes, I was led into a large physical therapy gym. I climbed onto a mat in the center of the room and was most intrigued by the row of sandbags arranged by size, some as small as beanbags and others as large as volleyballs.

More than forty years later, I would describe this memory to Miki Meeks, a producer at *This American Life*, as I participated in a radio interview between New York City and my Iowa Public Radio in my hometown of Des Moines.

"So wait, wait, just a minute," Miki said, laughter spilling into her sentence. "You've just been evaluated at the Mayo Clinic . . . possibly the best medical clinic in the world, and their advice to your parents is to 'sandbag the child'?"

I found Miki's humor contagious. For the first time in over forty years, I allowed myself the freedom to find humor and futility in the situation. I recall surveying the physical therapists encouraging me as I lay on the mat trying in vain to lift my sandbag-laden arms and legs and thinking *these people have no idea what they're doing*.

I wanted to return to Sue White. In her pool, I was important; I had great things to accomplish in my life. After being summoned from countless waiting rooms, I found a place in which I loved the sound of my own name. I wanted to work hard. I wanted to impress.

A few years later, my mom tried a more economical approach to swimming lessons. Immediately, I felt my stomach drop as we pulled into the drive, the hysteria of screaming preschoolers clearly audible beyond the chain-link fence. Instinctively, I understood these weren't private twenty-minute sessions with a heralded swimming instructor, but I wasn't prepared for the cumbersome pole of a pool net to be passed in my direction. To my horror, this swimming instructor insisted that I alone had to hold on to the pole as I demonstrated the crawl across the deep end. My carefully choreographed crawl disintegrated into a dog paddle.

"We don't want you to get hurt, dear," the woman explained as I made a quick exit from my lesson.

The warmth of entering the hot sauna of our station wagon fueled my temper. I scowled and crossed my arms against my chest, my angst clearly visible in my mother's rearview mirror.

"I want to go back to Sue White's," I growled.

With my return to swimming lessons with Sue White, it wasn't long before I performed up to her challenge. I swam the length of the pool as she walked along the edge.

"Stroke! Stroke! Stroke!" she belted out. I was up to the challenge, flipping like a dolphin in the deep end and making the journey back to the shallow end.

For the remainder of the summer of 1984, my nine-year-old ego soared with pride. As we crested the highway on our way to go boating in Saylorville Lake, the Olympic-size pool of Camp Dodge, which is the training grounds of the local National Guard, filled the valley below. The picturesque view of the giant pool filled with hundreds of bodies reminded me of the game my dad played with us. He'd fill a bowl with water and sprinkle pepper in it to represent swimmers. With a finger dipped in dishwashing liquid, he'd put his finger into the bowl, and the swimmers (pepper) would race to the edge of the pool. It never ceased to entertain us.

Later, at the pool's edge, I couldn't resist the wish to be a part of it all, not just the shallow end littered with preschoolers and mothers in conversation. I wanted the bravado of the deep end. As my mother was preoccupied unpacking our swimming gear, I slipped quietly toward the pool's edge and launched myself into the deep end.

Instantly, my rush of excitement stifled any fear that would have stopped me. I opened with a crawl, but I realized I would tire too quickly. I turned onto my back, listening to the menagerie of shouting voices and flailing limbs about me. With my eyes to the sky, I couldn't see where I was going, and I bumped into body after body.

Suddenly, one voice stood out over all the others. It was my mother! She had counted her brood and came up one child short and then spotted me in my orange swimsuit like an empty Cheetos wrapper bobbing in the sweltering August sun. I was out of hearing distance of the lifeguard, but I like to imagine he nonchalantly twirled his whistle on his pointer finger and said something like, "She's holding her own."

Minutes later, I reached the large raft in the center of the deep end. I tugged my beleaguered self ashore and fell open like a marooned starfish on the outdoor carpet. Bodies leapt about me sprinkling water on my face as my chest pumped excitedly for air. I had made it!

In addition to my sophisticated swimming lessons, my mom enrolled me in Mardella Keats's art studio. Just a quarter mile from our home, the retired art teacher had transformed her basement into a world similar to the low-lying thresholds in *The Hobbit*, with interior tunnels leading to tiny studio-size portals lit by study lamps and students perching on stools huddled over watercolor prints and charcoal drawings. I couldn't decide what excited me most—the jangle of car keys as the teenagers frequenting the studio left for their cars; the strong scent of Aqua Net sprayed to set a newly created masterpiece; or the earthen, damp bags of clay softened by the kneading of Mrs. Keats's hands. That I had a place here as a kindergartner instilled a wealth of confidence.

Soon, Mrs. Keats noticed I had a dreadful habit. As her footsteps approached, I would smear my hand to my charcoal drawing or smush my burgeoning clay sculpture into nothingness. I likened artistic mistakes to the trauma of my body: I wanted to start over.

"How about a turtle?" she asked one day.

Over the next hour, we fashioned a turtle basking on a stone. She circled the studio, keeping an eye on me, and occasionally came over to suggest tools for shaping my design and for me to select a color for the turtle's shell—I chose a vivid emerald green.

As I waited for my mother and younger siblings, I swung beneath the boughs of a low-lying weeping willow tree, my most earnest pumps tickled by low-lying fronds. There was magic in this place—swinging is the closest thing to running.

PRESENT DAY

As the 8mm film flashes before me, I see the purposeful look in my eye. My pelican legs propel sandal to pedal, and my head tilts back, blonde hair tossing side to side like an outtake from an Herbal Essences shampoo commercial. I'm too young to know I'm trying to be sexy, too young to understand that I can't change my world from the seat of my Big Wheel. In these archival frames of film, I squeeze my eyes closed in the warm rays of the sunshine and dream of an older me with tight-fitting gingham and skin like gingerbread, warm from the oven.

"You always did love the sun," remarks my mother.

"There's something else here," I insist as I reverse the reels and run the clip again, engrossed in the home-movie footage like an archaeologist on a long-anticipated dig.

For this second viewing, I scrutinized the timing of the hair toss. I mimic the tousled braids of Bo Derek in the film *10* just as my feet press the pedals. I witness the first glimpse of a rudimentary diagnosis and potential treatment: at seven years of age, I instinctively understand the growth and development of my body is stalled. If I can mimic the actions of a mature woman, I believe my body will spring to life. I'll sprout developed muscles, and my arms and legs will surge with renewed energy. Yet, even in my best attempts to pull ahead, Janet, two years my junior, laps me at every turn.

FALL 1980

A part of me was missing. I found it each Friday evening as the majestic strains of *Dallas* began and the *Dukes of Hazzard* left off. My favorite screenshot was that of a poolside Charlene Tilton, tossing back her long flowing hair. I strut about in my best attempt to transform myself into a svelte goddess with an eye for the camera.

There had been mention of my odd behavior in kindergarten. My teacher, Miss Van Dyke, lamented my signature hair toss as we'd settle on the carpet for reading time. The other girls in the class had begun to mimic me, especially my best friend, Gay. In photos, we were captured—linked arm in arm, hips thrust toward the camera, a hand combing back windswept hair, just like the models we revered.

Miss Van Dyke instructed us to draw pictures of what we wanted to be when we grew up. She nodded with the accuracy of a metronome as she moved about the room, admiring drawings of firemen, teachers, and pet shop owners. But, as she approached my drawing, the head bob stopped. She leaned in closely and repositioned her horn-rimmed glasses for a better view.

"What is this?" she asked with a wide-eyed snarl.

"I want to be a *Hee Haw* girl," I said.

Wordlessly, she turned my paper over. Her nostrils flared, and a bony finger pointed to the page.

"Try again," she retorted as she pursed her lips and surveyed the classroom.

"OK, grocery store checkout lady," I offered with a sigh.

I completed a new picture for my teacher, but I used boring colors—brown and gray. I drew myself without a smile, and I didn't add glitter.

At lunch, classmates asked me why my mom wasn't feeding me. Though I tried to push their hands away, someone's thumb and middle finger managed to make a loop about my upper arm or ankle.

"Look how small she is," they'd say, extending my embarrassment as a visual of my limb circumference was raised for all to see.

I was walled off and secretive even to those who loved me most. Though I was bestowed an idyllic childhood regarding things that could be controlled, I scurried toward loneliness like a hoarder of unspoken trauma.

"I hate this more than life itself," I would belt out in frustration when a puzzle piece didn't fit or my shoelaces became tangled in a knot I couldn't release.

Timidly, my mother would point out my faux pas.

"Don't you mean I love this more than life itself?" she would redirect, her facial expression twisted somewhere between a smile and a cringe. "Like you love puppies and ice cream?"

For me, it hadn't been a slip of the tongue. Anger percolated in me like the rattling of a teapot filled with boiling water. To my pastors and teachers, my many friends and classmates, I was the girl who always smiled, whose giggle rose above all others in a circle of play. Internally, though, I was plagued by an omnipresent poltergeist—one who grasped my shins to trip me or prompted falls that tore my white tights—a life I inhabited but didn't control. As I crossed the threshold of my home, I dropped my façade as my school bag hit the floor. I was tired of pushing down and covering up. I could no longer mop up my sadness.

The nape of my neck throbbed—a part of my body I came to understand as ground zero of my muscle disease. Mother bears instinctively take to this region to wrangle youngsters, a shawl of extra flesh and few nerves that is colloquially known as the scruff of the neck. For me, the region was far more sinister—the place where noose meets flesh. The genesis of my hairline at the nape of my neck surged with pain signals like a downed power line in a storm. When I asked my mother why it was that I hurt, she would explain it was my muscle disease, and I would abruptly clam up. Instead, I would seek solace in the bow of our magnolia tree, rubbing the flower petals, smooth as kittens' ears, until the hurt passed from me and flowed back into its natural roots, this Original Sin of mine.

At night, I would slip out of bed, often finding my mother mixing a batch of cookies or balancing her checkbook. Seated beside her, my feet planted on the chair dowel and nightgown drawn tightly over my knees, I'd remain silent for several minutes. She knew it took a while for the tears to come. Eventually, they would spill along with the hurts of the day; I had tried my hardest to overcome as my class exited the gymnasium door, secure in the knowledge that this would be the day that I could keep up as they ran. This time, I would force my legs to work, but instead, there was the burning of my leg muscles, the stabbing pain in my lungs, a bitter taste in my mouth. Then I was alone.

There was healing in just having a quiet place to share my troubles of the day. As the rest of the family slept soundly in bed, I cherished this time to have attention focused solely on me. My mother couldn't provide easy answers, but simply her presence, her listening, refilled the gas tank of my soul.

The call from my elementary school likely opened old wounds. I imagine my mother hurried to the school meeting with my challenges from preschool echoing in her mind—things can't go on like this . . . we need to make a change . . . surely, I must be falling again. But as she greeted my school principal, offering solutions for my walking difficulties, he shook his head and chuckled.

"No, no. That isn't why I called you in," he explained. "Jill's acquired her 'sea legs,' so to speak. Sure, she trips quite frequently when she's playing on the playground, but in the hallways, she walks in step with the other children. She's doing quite well."

As the principal and my mother paused in the kindergarten room doorway, they surveyed children sounding out letters and playing with modeling clay. I was observed engrossed in a book in the reading nook.

"Jill's reading *Charlotte's Web*," he said, "fluently!"

My mother had been aware of my fascination with *Charlotte's Web*, and specifically, Wilbur the Pig, for many months. I'd latched on to the book and developed a fascination for the "runt" size of many of life's choices—the smallest and most misshapen pumpkin in the pumpkin patch, the most matted stuffed animal in the toy bin at the toy store, and the most wilted flower that no amount of Miracle-Gro could resurrect.

My parents declined the request to advance me a grade, citing my small size as a deciding factor. Instead, I would join the first grade for math and reading each day. Ultimately, it wasn't without challenges, as I was greeted by whispers of "kindergarten baby" as I made my first journey across the hall.

I'll never forget the day my head separated from my body.

Of course, I don't mean this literally. As my head popped up after I surfaced from a dive, I couldn't help but notice the snickering of two boys just out of earshot. I assumed one was sharing a joke with the other or that maybe they'd opened a hilarious Bazooka Joe comic. Whatever the reason, I assumed it had nothing to do with me.

Again, I trekked along the narrow diving board. My gaze was fixed on the aquamarine waters below, sparks of sunlight glittering on the waves. As I prepared to take a dive, I sensed the muted stares of the two boys, huddled closely together along the chain-link fence. Their quietness irked me. I took a muted bounce off the springboard and disappeared beneath the waters.

Had I known what would face me as I surfaced, I might have stayed in this otherworldly rapture where bubbles of displaced air tickled my nose like the release of carbonation from a soda and my body turned somersaults and cartwheels, feats that were impossible for me on dry land. As I treaded water and took in gulps of air, it was unmistakable that something about me was the object of their hilarity. One boy whispered something to the other, and their bodies rocked in unison with frenzied laughter. I bypassed another jump off the board and joined my sister, Janet, as she was hastily collecting our swimming gear and stuffing it into a duffel bag.

"Why were they laughing at me?" I asked with pleading eyes as she polished off the last of her soda, the final sips of cola swirling in her straw.

"It's nothing," she said, her body backlit against the harsh rays of the summer sun.

"Tell me," I insisted.

"He said he'd never seen a white Ethiopian before."

The shock of her words took root in my torso, pumping through my bloodstream like venom, stifling my breath. I could have handled a put-down that could have pertained to any number of people—they didn't like the color of my swimming suit, my dive created too much splash, or my dive created too little splash. But the very personal nature of the jab left me reeling. The cruel depiction of human suffering—the emaciated bodies of starving children—likened to my body rendered me wordless and panicked. In the passage of a few minutes, my oasis from sadness—the unencumbered movements of my limbs in water—had become polluted indefinitely.

Though I didn't ask her, I wonder what my sister made of the painful expression on my face. Did she regret her honesty? Did she wish she had tempered her words with a nuanced retelling of events, one that would have stung less? For me, the experience of swimming in a public pool would never be the same. Every boisterous laugh, every eye turned in my direction, was a potential opportunity to be scorned and humiliated.

I was simultaneously intrigued and repulsed by the sight of a Barbie doll. I marveled at her flawless flesh and the perfect symmetry of her body and face, though I was frustrated by her rigid frame. Barbie's slender ankles, frozen in the contours of high-heeled shoes, couldn't balance her lanky body. Her tendency to tip required constant attention. Drawing the sleeve of a shirt, a miniature tube of fabric, over an arm frozen in flexion was an exercise in patience. Didn't she ever get the urge to stretch her arm? Barbie's neck was frozen in place, unable to bend chin to chest.

In some cases, Barbie's permanently flexed arms made sense. Stewardess Barbie required arms in constant motion, always willing to serve—arms at the ready to stow baggage, gather plastic cups, and position air-conditioning vents. Always poised. Always perky.

In time, my observations became more realistic. Who shows up for lifeguard duty in platform sandals? Exactly how did Barbie streak across the beach to save a life with toeless feet? Most curious of all was the sight of "New Mother Barbie," unable to look down and meet the eyes of the infant in her arms. Hadn't anyone at Mattel thought this through?

Sometimes a friend over to play would inquire about my Barbie collection. Kneeling before my toy closet, she'd open her own Barbie carrying case—a pristine white ensemble adorned with rose petals and a shiny gold latch. I'd witness a gasp of horror as I reached inside my Tupperware bucket and withdrew a seminude doll from a mass of intermingled limbs—stiletto heel wedged in a mass of tangled hair, three purse straps looped about her ankle. I didn't take good care of Barbie.

I never imagined the peculiarities of Barbie would someday be my own. As I progressed through first grade, I began to toe walk, not the adorable pirouette of a limber girl in a *Nutcracker* performance, but rather, a permanent deformity, as if my Achilles tendons had been mysteriously reabsorbed within my body. Physical therapists fought with the frozen joint, though my Achilles tendons were no more pliable than a wine cork. My feet, dipped in paint, formed a ghostly specter on paper—the balls of my feet and toes were visible but my heels and arch, absent.

At first, managing the tightness of my heel cords seemed like the best remedy, but by the age of seven, even a trip to the shoe store required advance planning. A half hour before our arrival, the code words "We're coming!" provided forewarning to the shoe salesman to reconfigure his merchandise. In a sweep of the store, he hid the most popular shoes, those decorated with lights and glitter and the most coveted of all—the Care Bear Collection.

He created a shrine of sorts to the saddle shoe, placing it center stage and at my eye level, and prayed that he could convince me that this very shoe (the one prescribed by my neurologist) was the hottest trend for rising second graders. In one lengthy visit, my heart was won over as I was introduced to saddle shoes with a pink stripe—a special order from a catalog.

Soon, the need for surgery became abundantly clear, as a trip to the mall found me lying on the floor to soothe my cramping calf muscles. As my mother scooped me up and carried me to the car, I didn't offer any protest. It was clear to both of us that if something wasn't done, and done soon, I would lose the ability to walk for the rest of my life.

Each exit interview from the Mayo Clinic was accompanied by a manila folder of physical therapy exercises. Inside, healthy children in black leotards stared toward the camera with expressionless faces while demonstrating stretching techniques. Sometimes bodiless hands entered the frame, hands with perfectly manicured nails holding arms and legs in unnatural positions. I hated these photos; they were the antithesis of me. I longed for a dance recital photo in my dad's wallet: me, all smiles and sequins, slipping onto a convenience store countertop as my dad reached for a twenty.

On the car ride home, I hid the manila folder, sliding it into the crack between the station wagon seats. Inevitably, the manila folder found its way home. Each summer evening, as the gentle hum of the cicadas heralded the approach of a new school year, I joined my dad on the chaise longue on our back porch, lying placidly as he prodded my frozen joints in ways they would never yield.

We never spoke during these encounters. I acquiesced, but the emotional hurt inside me was palpable to both of us. In these moments, he sensed the hollowness of my soul, the crumple of my body like a lifeless puppet. I was slipping away.

One evening, the manila folder disappeared and never returned. I hadn't moved it this time. Instead, my dad beckoned for me to join him on the chaise longue, paddle brush in hand. I savored the strum of the brush through my hair and the touch of his palm as it glided over the softened strands.

"You can count each stroke of the brush," he explained. "Each night, you can brush your hair one hundred times, and it will spill down your back like rays of sunshine."

My dad used to count the duration of my stretches. He never spoke aloud, but sometimes I'd catch a glimpse of his lips mouthing the count.

Now, we spoke in unison, counting the strokes to one hundred like a parent and a preschooler rehearsing a favorite nursery rhyme.

In these moments, I imagined I was my dad's wounded trophy horse. He knelt beside me in a darkened stable, drawing a brush through my coarse mane. The triumphant blare of the bugle echoed in the distance heralding the start of a race, but he stayed with me. Later, there was the clop of hooves, the strobe of the camera flash, and the enchanted dance of wayward petals over the wall of the stall. Petals, the broken promise of a flower, he resurrected and wove into my mane.

"Dad, who will love me?" I asked.

The brush paused midstroke, and I felt the slight tremble of his hand.

"You'll fall someday, and a rich man will pick you up," he explained.

It didn't matter what his response was. He spoke freely, without missing a beat.

"He'll brush your hair from your face, and he'll realize you're a beautiful girl."

"Oh, Dad," I exclaimed with a giggle.

"You think I'm joking, but you wait and see," he said, as he tapped the brush on the tip of my nose.

There were missed opportunities in my life, times the effect of "aloneness" could have been lessened. I recall extra PE sessions in elementary school when I was paired with a girl, several years my senior, who took to the gymnasium with loud slaps of the basketball. I trailed her with enthusiasm. She was a curiosity to me—her long blonde hair tossed wildly with each step while her hips moved in an exaggerated manner and her backside stuck out. Did she want attention? Had no one told her it was impolite to walk in this manner?

Little did I understand, at the time, that I walked with the same curious gait—my shoulders tipped behind me, tiny stomach pooched forward, and pelvis tilted backward like a woman in high-heel shoes. Like most people with muscular dystrophy, the muscles that anchor the body vertically, the hip flexors, are weak; as a result, they walk with the same unusual posturing—what doctors describe as lordosis. At the time, I didn't make the connection. No teacher explained to us that we shared the same medical diagnosis. Perhaps, she, too, followed behind me and pondered my odd posturing. Maybe she thought I was a show-off, someone clamoring for the spotlight.

I wouldn't understand until adulthood that this unwanted pelvic tilt held paradoxical charm. Twenty years later, as I watched my friend Mary, who shares my diagnosis of EDMD, being pulled from her motorized scooter into a standing position by her husband, I couldn't help but notice how her increasing disability had advanced the curve of her lower back. There was an enigmatic attractiveness in the way the disease affected her body, though I couldn't quite put my finger on it.

"It increases my femininity," a young Finnish woman with EDMD once explained to me.

For me, this ethereal attractive quality was captured on film as my sister fanned out my wedding gown. The corset stays and laces down my back added twenty pounds to my figure. I could barely recognize myself when framed in this flattering gown. Throughout the ceremony, I clung to Jeremy's shoulders and leaned into him for support. I was, perhaps, the only bride to cling to her husband for the entirety of a wedding ceremony.

Ultimately, it would be this advancing pelvic tilt that would culminate in my loss of the ability to walk or stand upright by the age of thirty-three. Throughout my twenties, I could use a heavy book bag to alter my center of gravity, holding the bag against my backside to push my weakened hip flexor muscles forward. In time, the strategy required more and more weight, and finally, there was no amount of weight available to counter the advance of my muscle weakness.

My mother met me at the back door, bra in hand. It was the third time this week I had tried to go "T-shirt commando," and the look in her eye signaled I was not going to win this time. My breast buds strained against the thin cotton fabric like twin corks, and I crossed my arms to cover myself, glaring back at her with fiery eyes and unkempt hair.

The carpool arrived in our driveway. The bleat of a car horn pierced the silence of a still spring morning.

"You have to wear it," she insisted, handing the bra to me.

This had been my idea just a few weeks ago. As my mom prepared dinner, I had sat at the kitchen table, stirring a bowl of waxed fruit for nearly twenty minutes before working up the courage to ask the most important of fourth-grade questions. I wanted a bra . . . needed a bra . . . all the other girls had one.

We had perused the lingerie department of our shopping mall together, selecting training bras with delicate lace filigree and satin ribbons. I had spun a single rotation before the dressing room mirror, admiring my mere whisper of a svelte form.

On this day, I had refused to comb my hair. I spewed angry teenager speak, erupting at the slightest provocation hoping that someone would find a hurting me beneath it all.

The truth was, my pen pal had rejected me. We'd exchanged letters for several months as part of a social networking project with a neighboring elementary school. Initially, we'd gushed with the sharing of our lives, even adopting a secret cuneiform code to translate our deepest secrets. At the big reveal party at the close of fourth grade, she snubbed me as it was revealed that I was her match and she saw me face-to-face for the first time.

"My pen pal is totally weird," she'd giggled to her friends as she surveyed my bird legs.

I hadn't even thought about mentioning my weak muscles. It was something I'd always considered a private secret I pondered within myself. In the space of a few weeks, my appearance had become an oddity, the onset of puberty fueling a wider and wider digression from normalcy. What came to change me, came to maim me.

I didn't know how to express this deep hurt, even to my mother. Reluctantly, I grabbed the undergarment from her, changed in the bathroom, and joined the carpool, silent and brooding.

I sat on my hands and resisted the urge to reach out toward the glass custard dish. It glided across the Ouija board, settling on numbers and letters, as the slumber-party girls leaned in for closer inspection. My arms felt oddly electrified, and I knew better than to reach for the glass. With eyes closed, I imagined my arm reaching out and my fingers brushing the glass, causing it to spin feverishly and strike the wall, shattering into thousands of shards.

In confirmation class, we nonchalantly asked if Ouija boards were real. I'd seen them on department store shelves, tucked alongside innocuous board games such as Operation and Trivial Pursuit. We'd expected our Pastor Litzner to dismiss Ouija boards as a phenomenon of hysteria, a relic of less-developed minds, but his words were unambiguous.

"Never touch a Ouija board," he admonished. And so we never did.

I'd checked out all the books on ghosts and poltergeists at my school library, and some, I'd checked out twice. If I could believe a candlestick could hurl across a room or a bed could levitate, perhaps one could believe the impossible in me—it's my body that's haunted.

I could hear my dad shouting from the kitchen. He'd found my books that had spilled from my book bag.

"A child's library?" he bellowed, slamming one of my books on the table for emphasis. "I've got a mind to go in there and give that librarian a piece of my mind."

I could understand his fear. At first glance, it appeared I'd developed a fascination with the occult, but this wasn't the case.

Studying the Bible at church, we read of Jesus's miraculous healings, and I was fascinated with the passages describing people possessed by demons. We were taught in confirmation class that these were stories likely invented by those not yet informed of modern medical practices. There was something I understood, related to, in a way that others didn't. I found enlightenment where others found ignorance. I imagined a palm of blessing pressed tightly to my head and a figure stirring within me destined to be set free.

In a photo taken of me at the age of ten, I am seated on a picnic blanket with my dad at the beach. Like my dad, both my arms are locked in flexion by approximately forty degrees. I had never noticed the oddity in my dad even though it was right in front of me all my life, and I don't have any memory of sensing that my arms were becoming like his. There's something beautifully bonding about us leaning toward one another, the same arm passing from one generation to another.

By early adolescence, new symptoms emerged that were impossible to dismiss. It began with light taps to my spine as my piano teacher circled behind me—an ear to the piano and an eye on my curious posture.

"Relax," she'd insist with each tap. "Let the music flow through you."

I had dreamed of continuing with piano performance long past my elementary school lessons, but as the taps increased in frequency, I knew my dream was coming to an end. I had studied concert pianists and noted the fluid movements of the upper body were as integral to an evocative performance as fingers dancing over piano keys.

By the start of sixth grade, new classmates began asking about my car accident. In bewilderment, I explained that I had never been in an auto accident nor had whiplash. I was aware that something was terribly wrong with my spine—seemingly overnight my muscles had tightened as if concrete were poured into my spinal column and hardened. My ability to turn my head to the right or left was reduced to a few degrees in either direction. I had no ability to lower my head to my chest. In church, my mom prompted me to sit back in the pew, yet my spinal rigidity allowed for only two postures—fully upright or lying down. Besides the obvious limitations of my predicament, my nearly overnight transformation clouded my social interactions. My dad shared minimal neck and spine rigidity, but overall, he appeared as a distinguished and physically well man.

After school, I lie prone on my bed with my head and neck off the edge of my mattress. I draped the strap of my school bag onto the nape of my neck, but no amount of counterweight could stretch the frozen joint.

At the close of my sixth-grade year, a curious change took place. I'd once loved my bike, but my attempts to ride had become disastrous. My arms quivered as I tried to lean my torso toward the handlebar. I fell forward forcefully, and an angry bruise filled my chest.

Later, as I invited a friend to the roller rink, I occupied the snack bar the entire visit. I divided my attention between waving to my friend as she looped past and studying the rotation of hot dogs in the warmer. I couldn't help but recall I'd skated well only weeks earlier.

My friend Dawn and I dressed as "fifties girls" for trick-or-treating on Halloween. Though we'd only made cryptic plans on the bus ride home, she arrived in red and I in blue. I couldn't help but feel the magic of the night, as if we'd stepped out of a time machine. We spun like twin teacups, the crisp autumn breeze filling our poodle skirts, the silk ribbons in our hair billowing behind us like the tails of kites.

I joined Dawn for the first few stops, excited to try out new jokes and fill our plastic jack-o'-lanterns with candy. Within a few hundred yards, my legs quivered with exhaustion. Jokingly, I asked Dawn to fill a "to-go" order for me, and I sought refuge in a grove of trees, out of view. My fingers dug into the bark as I struggled to hold myself upright. A foreign stirring sprouted within me, like an outstretched hand wafting back and forth, then the staccato of fingernails strumming flesh low and deep in my abdomen. A car stereo, cranked to full volume, blasted the eerie strains of "Welcome to the Jungle," the new release by Guns N' Roses, into the starless night.

Though we completed our journey just blocks from where we'd begun, I called my dad for a ride home. Upon our return, the staircase loomed before me. I struggled with each step and paused midway to rest my torso on the railing. From this position, I saw my dad watching from the base of the stairs with a look of concern on his face. We lived in a two-story house, and we each climbed the staircase multiple times a day. We were not in the custom of watching one another ascend the steps.

"Go get Mom," I insisted.

She found me, moments later, in the second-floor bathroom. Without words, she understood what had taken place by the look on my face. She knelt beside the tub basin and drew my face to hers.

"You understand what this means?" she said. "You can be a mother."

I was not sure what to make of these words. References to me as a mother had all but disappeared from conversation. When I was younger, dolls were placed in my arms with the ease of an answer to a question. In the context of recent changes to my body, the conversation seemed foreign. Something about this night was very wrong. Growth spurts fueled my fire—my muscle weakness—but I couldn't find the words to make her understand. On this night, I was doused in gasoline.

Above the cacophony of sizzling burgers, fryer requests, and dropped cookware, I heard the gasp, "Oh, God!" and I knew the words were meant for me. As I continued to sculpt the ice cream cone I held in my hands, I'd hoped the perfect soft serve twist would satiate the lunch patrons in my line. Reluctantly, I turned to face them and presented the lackluster dessert, ominously tilted to the side.

"Whatever could have happened to you?" the elderly woman cried. "It's your back, dear. Your back is caving over."

Though her words were cryptic, I knew I had scoliosis. The customers in the line behind her grew impatient, but still she stayed. Not knowing what to say, I dumped the melting cone upside down in a cup and wiped the rivulets of ice cream with a napkin. Her husband placed a few bills on the counter and refused the offer of change.

They selected a booth close to the cashier line. Engaged in animated conversation, they turned back to look at me after a few moments. The dessert remained untouched on the table. The ice cream melted, barely contained in the cup.

In the few moments my workstation was free, I escaped to the restroom. With my back adjacent to the wall mirror, I gazed at my reflection in the compact mirror I held in my hand. The gray, vertical lines of my McDonald's uniform distorted into angry bent lines. My orange necktie sat cockeyed. My right shoulder jutted a couple inches higher than the left, fixed in an unorthodox position as if jolted by electricity; the misshapen contours of my back were cringeworthy. As I blinked back tears, I returned to my workstation, glancing periodically at my reflection in the chrome of the coffee machine.

Only months ago, I had passed the scoliosis test at my junior high school. I waited uneasily in a line of girls from my PE class, certain I would be one of the few summoned to the coach's office. One by one, we were asked to remove our shirts and lean forward and touch our toes. The nurse examined the back of each student, dismissing most to the practice gym. For some, her finger traced the arc of the spine—a grimace on the face of the girl as she realized the exam was taking too long. My check was brief and uninspired.

Brushing away tears, I returned to my cashier post. There were real crises to solve here—someone's burger had too much ketchup, and a boy had been given a girl's Happy Meal toy. As the buzz of activity slowed, I leaned forward, placing my elbows on the counter and looked out into the dining room.

"If you've got time to lean, you've got time to clean," the manager said as he rounded the corner, sliding a towel and disinfectant spray in my direction.

"So how are *we* today?" the nurse inquired as she slid my X-ray into the lighted box on the exam room wall. Too startled to speak, the otherworldly image hovered before me—my spine, a serpent in water, smooth curves immersed in a shallow dark pool. It was the first time I'd had the chance to gaze upon the monster inside me.

"Remove your clothes, except your underpants," said the nurse as she explained the features of the gown with the charm of a flight attendant. "Don't forget, open in the back."

I was half changed in the corner cubicle when I heard the obligatory knock. I counted six pairs of leather loafers beneath my changing room door, two with tassels, four without.

"We're dealing with a very serious situation," suggested a single authoritative voice.

The ominous words wafted over my changing room door. Teetering past the crowd while holding my gown shut, I was ushered toward the doctor seated on a stool. He introduced the young doctors en masse. The senior doctor directed me to turn my back toward him, then lean over and touch my toes.

"May I adjust this just a bit?" asked the doctor.

I simply shrugged my shoulders as he lowered my underwear a couple inches.

"You'll notice the winging of the scapula and the compensatory curvature of the lumbar spine."

A timid knock on the door interrupted the presentation, and a seventh pair of loafers joined the observing doctors. No tassels this time.

"This is a severe deformity of the spine secondary to neuromuscular disease," explained Dr. Klaussen.

Chapter 3

I pressed my thumbnail to my shin creating a crescent-shaped indentation, words muffled like a conversation heard from below water. Carving deeper, I drew blood.

"In comparison to the previous X-ray, this curvature has progressed nearly twenty-five degrees in a short period of time," explained the surgeon after he'd excused me to the exam table.

"What about a brace or physical therapy?" my mother asked.

I cringed at the mention of a brace, but my fears were short-lived. My condition was far too severe. The surgeon and my mother volleyed questions over my head.

"What would happen if she does not have the surgery?"

"She will die," said Dr. Klaussen without hesitation, "from a combination of heart failure and suffocation."

I wondered if I was the only person in the room to allow this revelation to pierce the psyche. In any other generation, in many parts of the world, I would not be able to access such complex and expensive medical care and would not survive past my fifteenth year. The thought filled me with a mix of sadness and curiosity, as if tarot cards had been laid out before me whether I chose to see them or not.

Dr. Klaussen and a circle of residents swarmed about my X-ray. Alone, I felt the crush of my chest and my slow, stifled breath. Death no longer remained an obscure concept. It had a movement, a force, a feeling that was tangible. This team of surgeons can save me, but I wonder about the girl who will continue my life. Will she live as a ghost, invading a life destined to end at the age of fifteen? I studied the ominous silhouette of my spine and considered one last coil, one final snuff of breath.

My mother perused her checkbook calendar, proposing potential surgery dates with the ease of a PTA member planning an after-school mixer. Unattended on the exam table, I seized the opportunity to sift through my medical folder. One entry indicated I have "juvenile, proximal muscular dystrophy," while another entry suggested I had a "myopathy of unknown etiology." And yet another indicated a diagnosis of "limb-girdle muscular dystrophy." Despite the incongruity, the word *dysphoric* was used to describe me on many occasions.

I mulled over the word *dysphoric* in my mind before I asked the meaning of the word from the sole intern to make eye contact with me.

He appeared startled by my question. No one but my mother or Dr. Klaussen had spoken aloud in the last fifteen minutes.

"Profoundly sad," he said quietly.

My tear-stained face dampened the mood for polite conversation. Offering a consolatory smile, the nurse led me into the pediatric neurology department and opted for the hallway to the right. In my ten years of visits to the Mayo Clinic, I had only ever gone left; my curiosity was piqued. She opened an unmarked door and leaned against the doorframe, arms folded, as she spoke to a passing nurse.

"I think Earl's going to take me to a movie tonight," she said. "*Glory*, I suspect."

Through the open door, I caught a glimpse of the room—wood-paneled walls and green shag carpet—reminiscent of an office rather than an exam room. Dim light illuminated the curious space within. Perhaps I might find a sofa to lie on and a counselor in a wingback chair eager to listen to me.

"Um-um," the nurse continued, "you know it. That Denzel is mighty fine."

How curious. Ducking beneath the nurse's arm, I explored the room on my own. It resembled the dressing room of a high school drama department—large naked bulbs illuminated a vanity mirror, and a pile of slippers were tossed haphazardly in the corner. Contents of a lost-and-found box spilled over the sides.

Like the garbled voice in the *Peanuts* cartoon, the nurse offered instructions. She mentioned something about putting my hair up; then she grabbed a garment from a bureau drawer and placed it on the vanity table. I examined the white cotton fabric, which had no discernable front or back.

"Where's the front?" I asked as she exited, closing the door behind her. The room once again grew quiet and dim.

For a few moments, I paced like a cornered tiger until a closet drew my attention. Sleeves of various colors had escaped, begging for perusal. With a gentle tug, the accordion-style door gave way to reveal robes for men, women, and children. One, a child-size robe, was decorated with rainbows and unicorns, another with rockets propelled through a constellation of stars.

Like the star of a sci-fi horror movie, searching the charred remains of a planet for signs of previous inhabitants, I moved about, communing with those who had passed this space before me. Perhaps this was the only reason I mindlessly disrobed and pulled on the awkward garment.

Topless, I took a seat on the floor and sorted through the lost-and-found box. Baby bites nipped a rattle handle, maybe a remnant of baby teeth already lost and replaced. A young woman appeared perfectly healthy in her driver's license photo. She designated herself an organ donor and indicated Y for medical alert. What lay beyond the bounds of her photograph? What circumstances had led her to this mysterious room?

Curiosity prodded me to move about the room once more, and the slight touch of a wooden door sent me toppling into an adjacent dark room. Ashamed of my nakedness, I crawled back toward the changing room.

"I was about to come and get you," a woman's voice beckoned. Obscured by a large camera, her voice felt bodiless. Two tennis shoes perched atop a footstool.

As I ventured from my hiding place, I took in my surroundings. The room appeared like any other photography studio, complete with bright lights, background screens, and a sheer umbrella hanging from the ceiling.

The meager selection of scrolled screens left few choices—absent the image of a roaring fireplace complete with stockings and ivy or billowy clouds upon a blue sky background. No winding walking trails enveloped by crisp autumn leaves. A red wagon wouldn't be rolled on the set or a carpet-covered bolster for positioning. Instead, dull gray, the only choice of color, and a yellow life-size ruler, reminiscent of a police lineup photo, provided the only decorative touch.

She brushed my hair and cast my jewelry on the table, then twisted my hair into a bun high atop my head. Circling my frame, she eyed her work with the care of a mother preparing her daughter for a dance.

My hair was my protection. Baby fine and grown nearly to my waist, it enveloped my body like a giant silk scarf draped over a wilted pistil.

I can't hide from this experience, though it will haunt me. I will wonder why I didn't fight, why I didn't run away, why I allowed this to be seared into my memory.

Closing my eyes, her nimble fingers massage my scalp as if she is weaving bits of baby's breath in my upswept curls. An imaginary arm hugs my waist. A corsage kisses my wrist; I am drawn close by my imaginary love. Parents duck among one another, positioning cameras and prompting smiles.

The fantasy could last only so long. Like Cinderella missing her curtain call, my true identity invades the photoshoot. The strap of my sequined gown slides off my slumped shoulder, and my concave torso tilts to the right, flesh spilling over the side of my gown like warm dough pooled against the rim of a muffin tin.

If this took place at school and "normal" students experienced this voyeurism, no one would stand for it. Within minutes, angry parents would arrive on scene, the principal's office would be breached, and competing news vans would streak across the campus lawn. I knew better. This urgency wasn't for me. A doctor had ordered these photos. This was my new normal.

The photographer reminded me that the photos are part of my scoliosis file. I'll return in several weeks for "after photos."

Will one doctor or a team study this photo? Does it matter? Will this humiliating image be tucked away in my medical file or banished to the darkened tomes of a hospital basement? Will my name and face accompany this record, or will this eerie day be classified by a number?

A medical release form was a foreign concept to me. My parents signed dozens of forms as I sulked in the waiting room. They were advised of complications of surgery, the postsurgical follow-up care. Perhaps a photography release was buried somewhere in this cumbersome paperwork.

Cards had begun to arrive in the mail. Kittens danced with butterflies. Drops of dew slid down rose petals. God is with you, the verses reminded me, but there was no card for this place, for these feelings.

A plastic crate, filled with stuffed animals, appeared out of place beside the monstrous camera. Babies came here. This reality was a concept my mind was reticent to accept. To summon a baby's innate joyfulness and trust to secure medical documentation did not sit right with me. Perhaps a parent watched nearby, blinking back tears, as they contemplated photos no one would want to order.

Did anyone cry before the camera? Did the photographer offer a tissue? Crop out the face? How did the photographer survive in a room where there was no healing, only documentation? How did she arrive at such a position? An application? An interview? Had she began her career at JCPenney, taking photos of children? Alone, in the studio, did she pass before the lens of the camera, or did she slink along the periphery, sidestepping the camera's aim?

I had seen these photos before—a stark, frozen moment of a patient's greatest vulnerability, the body positioned in a way nature and the photographer dictate, all except for the eyes. The eyes cannot be manipulated or coaxed. It is often said that the eyes are the windows to the soul. Maybe that is why black bars are printed over the eyes of the patient. Perhaps this is done to protect the patient's anonymity, but I wonder if it isn't really done to shield the peering eyes of the medical community from the humanity before them.

A pair of footprints painted on the cement floor defined the camera's aim. Like a newbie detective stepping around the chalk outline of a murder victim, the photographer circumnavigated the feet. I froze in place, my body conforming to the macabre scene like a department store mannequin. The photographer tucked some hair behind my ears and opened my folded arms and placed them to my sides.

Maybe my photo will appear in a medical text, viewed by a sleep-deprived intern. He will think of his daughter, a girl so different from the patient with the hidden eyes in the bleak photo. He will return home, greeted by his shrieking toddler, spaghetti tossed on the wall beside her high chair. In these moments, he will cherish her fight, her strong will, and even her tantrums. She could never be like this girl who stands motionless before the camera, this girl who is surely intellectually dulled. A vision of a four-year-old me—one kicking and bucking to fight Velcro restraints—invaded my soul. I missed her fight, her screams, this nearly forgotten echo of me, this girl I only knew in the third person.

"Honey, I'm going to need you to lower your arms."

Despite the harsh lights, I shivered, my folded forearms clasped against my breasts.

For the first time in years, I did not do what I was told. My resistance was resolute. My perspiring skin glistened beneath the hot lights, and the fine, downy hairs on my arms raised in defiance. Seconds ticked past. The diaper-like garment slipped past my narrow hip bones revealing pubic hair. Momentarily, the photographer joined me before the camera lens, like prey in the scope of a rifle.

"I hate you," I hissed.

Wordlessly, she positioned my arms at my sides and disappeared behind the camera. My soul watched this ghost of me from the refuge of the studio rafters. The camera clicked. I was captured.

4

LOSS OF CONTROL

"No one faint," admonished the nurse to the patients seated at the kitchen table.

It was 1996, and my family had gathered for underground genetic testing. The OSHA violations were too numerous to count—there was no laboratory paperwork, no consenting physician, and no protocol for testing. The phlebotomy supplies had been lifted from a hospital and a nurse cajoled into participation. Then there was me, the twenty-one-year-old college student, with muscular dystrophy, who managed the covert operation. I shrugged my shoulders when asked, "Heparin or EDTA?" Clearly, I was in over my head.

Playing simultaneous roles of patient, scientist, and physician, I filled a shoebox with packing peanuts and eight slender vials of blood, one from me and seven drawn from family members—five affected and two controls. Dr. Daniela Toniolo, from her laboratory in Pavia, Italy, requested my dad's half brother, Dan, be included as one of the controls, and he graciously accepted.

The following day, I safeguarded my family's legacy with packaging tape and permanent marker and stepped to the window of my local post office.

"Blood," I said, a bit too loudly as I was asked to declare the contents of my international parcel.

My answer spurred curious glances, but for the postal clerk, my odd declaration barely drew a blink and a nod. Drawing a binder from beneath the counter, he ran his finger along accepted shipping parcels for Italy and granted permission for my package to ship overseas.

The traditional Sunday dinners of my youth adhered to a predictable pattern: Dinner consisted of roast beef, mashed potatoes, and baby carrots. The meal would be served no earlier than two hours past noon, and my Uncle Dan would ask for a Diet Coke the moment my mother took a seat at the table. He didn't intend to pester her. An empty glass signaled a Pavlovian response: it was simply time for a refill.

Following supper and a third request for Diet Coke, my mother's urgent eyes signaled I was to occupy my uncle for a few moments in the adjoining room. I'd feigned interest in baseball stats, nodding with glazed over eyes, and dropping "uh-huhs," and "reallys" in the few lulls in the one-sided conversation. Sadly, I never really connected with my uncle.

The truth was that Uncle Dan was a good person. From my earliest memories, he arrived with fragrances on my birthdays. When I was a young child, he gave me Tinkerbell perfume with a lavender atomizer; a few years later, Wind Song spray, with the advertising jingle "Your Wind Song stays on my mind," the caption still fresh after twenty years; and in my late teens, enough Jean Naté Bath Splash to swim in.

His thoughtfulness was bestowed on my mother as well. Uncle Dan would arrive minutes before supper with gifts for the occasion—"virgin" carrots, complete with leafy greens and taproots, or a frozen ham "to toss in the oven." At Christmas, he was sure to bear a poinsettia, my mother's catlike reflexes rescuing an antique trunk or piano bench from the dribbling planter.

My dad confessed that he experienced frustration with Dan, particularly in his college years. Uncle Dan would seek out my dad at Drake University, crossing the campus like a frolicking puppy run amok. From a distance, he appeared "normal," but as he joined in, the circle of conversation would fall away. My dad would silently lament as to why Dan couldn't be just a few decibels restrained in conversation, why he couldn't interpret vocal cues and facial expressions, why he had to be so different.

I asked God why it was that an accident, so early in his life, could have altered my uncle's experiences forever, why communicating with others had to be such a struggle when he clearly drew satisfaction in sharing what was inside him. I'd come to understand this frustration was a two-way street—the wish for a more normalized life was Dan's as well.

"I want to be the sort of person who carries important papers in a briefcase," he once remarked to his mother. "That would make me truly happy."

I loved the way Uncle Dan could articulate his greatest wish for happiness so succinctly. It reminded me of Dorothy, the Tin Man, the Scarecrow, and the Lion gathered about the Wizard of Oz to receive their greatest wishes in the form of trinkets.

Perhaps my dad pushed Dan away because he had his own battles to fight. In early childhood, my dad's sluggish run was attributed to a brush with polio. The hurt of being picked last for team sports left little patience for Dan. At all costs, he had to distance himself from disability, particularly as he rose in the ranks of the ROTC and dreamed of a career in the Air Force Academy. When a myopathic muscle biopsy derailed his career in flight, my dad pursued a career as a federal prosecutor. Though he walked with a waddle, he adopted a lurch, keeping pace with those about him and blending in almost seamlessly like a survivor of a war injury. It wasn't until my preschool teacher called, noting in her conversation that my father and I walked with the same unusual gait, which my parents acknowledged was a hereditary condition.

On weekends, my dad entertained Uncle Dan, often with visits to Iowa Cubs' baseball games. As we traversed the stadium, Uncle Dan's gigantic calves drew curious stares from onlookers and "anabolic groupies," those following in the hopes of gleaning weightlifting tips.

At times, Uncle Dan would dawdle far behind, and we'd turn back to see him conversing with a credit card solicitor as he perused a table of Frisbees and cotton T-shirts. My dad would usher Dan from the table with the admonition that he'd never be approved for a credit card.

"Well, I know that, and you know that, but he doesn't know that, and I get a new cap," Dan replied.

I masked a sheepish grin. Uncle Dan was truthful to the core and remarkably resourceful when he needed to be.

In 2025, the entire human genome can be analyzed in a matter of hours. In the mid-nineties, however, the search for a disease-causing genetic variant could meander for years, requiring the assistance of dozens of scientists. Finally, in 1999, the results of my journey for answers arrived in my email inbox. My fingers hovered over the mouse as I realized the gravity of my situation—there was no doctor, no nurse, no genetic counselor. I paused momentarily, and then, with a gentle click of the mouse, I unlocked my genetic code.[1]

A single *G* had been changed to a *C* out of the three billion letters in the human genome. Though most individuals possess a handful of innocuous misspellings, these variations were seen in the same gene, nuclear lamin, in all five families with Emery-Dreifuss muscular dystrophy. We'd finally found our gene!

My family's results were as expected. My father, sister, and two brothers shared my genetic variant, and my mother and healthy sister, Janet, were confirmed as controls, but a most curious result startled me. My father's half brother, Dan, a control for the study, also shared the genetic variant. I was bewildered. My Uncle Dan—a man with bulging muscles and a broad grin to rival "The Fonz" from *Happy Days*, just the kind of person one seeks to move a heavy sofa—had inherited the genetic variant as well?

Days later, my family gathered around the same kitchen where we'd originally drawn blood, and I shared the results of the genetic study. Uncle Dan's results were met with silence.

"They sure messed up those test tube labels," I offered, to raucous laughter.

It never occurred to us that we had been presented with the truth. We were simply a family pondering a medical mystery, unaware we may have stumbled upon one of the greatest finds in the field of muscle biology. In the years that followed, we would find there were dozens of cases from around the world of family members assigned as controls for genetic study who were found to share the same genetic variation as a severely affected family member (i.e., one appeared asymptomatic and the other required a wheelchair).

As the years passed, the Sunday routine remained the same. Uncle Dan's predictable Diet Coke request would come just as my mother sat down at the dining table, but this time, there was a subtle change. As my uncle extended his empty glass for a refill, his hand tremored. Perhaps I would never have noticed except for the ice clinking in his glass.

Weeks later, Uncle Dan rolled out of my parents' backyard pool like a bloated carp.

My family surrounded my uncle as he lay motionless on the concrete. Something was terribly wrong. Numerous tests with the VA hospital yielded no answers. Uncle Dan was very ill, but no trace of cancer or infection could be found. His sudden illness was a mystery.

At the close of the summer, my sister, Janet, entered our family's surname in the Great Iowa Treasure Hunt, at the Iowa State Fair. The exhibit of computer terminals was intended to match long-lost funds, such as unclaimed payroll and tax refunds, with the proper recipients. Janet had not expected to find any results pertaining to the family, but the search did reveal unclaimed proceeds of a life insurance policy purchased by my late grandmother.

The thousands offered to Uncle Dan were nothing short of a windfall. A handful of coins discovered in his sofa cushions or the ashtray of his car sometimes sustained him in the days leading up to payday. Perhaps these unexpected funds could help him as he struggled with his enigmatic illness. Instead, my father gasped in shock as he reached inside his birthday gift wrappings and felt the velvety texture of fine leather.

"It's for you, Bob," Uncle Dan explained. "It's to carry around all your important papers."

We were overcome with love for Dan in these final months. Though we heaped his plate with meat and vegetables, Uncle Dan was imploding metabolically, the muscle and fat of his body seemingly melting away. The greatest loss was the ebbing of conversation—Dan's once excitable love of baseball was absent from the meal. As he rose from my Thanksgiving dinner and waved goodbye, I feared that I would never see him again. He was only sixty-six at the time.

My uncle fell on a particularly blustery winter day in the parking lot of his apartment. A passerby had witnessed the fall—his legs slipping out from under him and his head striking the concrete. He'd urged my uncle to let him summon an ambulance or drive him to a walk-in medical clinic, but my uncle refused the offer of help and insisted he just wanted to lie down and rest.

Uncle Dan never woke from his sleep and passed away from bleeding of the brain. Following his funeral, I spoke with his daughter, Krys, my cousin. She mentioned weak legs and his tragic fall. I couldn't believe I missed what was right in front of me. I'd been so overwhelmed by the shocking and sudden loss of mass from his body that I hadn't even considered muscle weakness as a factor. Then it came back to me: the hand tremor and his accepting an outstretched arm as he shuffled to his car. Perhaps Uncle Dan's genetic test results were correct but were too unbelievable to take in at the time.

I've often wondered what happens to the concerns of this world when we pass from this life into the next. I sincerely hope we get a chance to relive the moment we chose a wrong fork in the road and set things right with those we have hurt.

My moment is abundantly clear to me: 1984. My parents' station wagon backs slowly out of the driveway as my Uncle Dan waves from the curb. We turn due north toward the Mayo Clinic, to the place with all the answers, doctors at the ready to solve our genetic mystery, this mecca of medical knowledge. If given the chance to live this over, I would choose this moment to leap from the back seat, sprinting on my new legs, and throw my arms around my uncle and his briefcase of important papers. My Uncle Dan—the first person I run to in heaven.

5

KILLER INSTINCT

I've always been attracted to true crime shows. Despite the macabre nature of the genre, I love being drawn into a complex web of characters, munching on popcorn as I decipher forensic evidence and make predictions of likely suspects. As an armchair detective, I enjoy predicting the moment a casual brush with a stranger strikes a malevolent chord and imagining how I would sense and escape from a vulnerable situation.

One afternoon, I found myself drawn into an episode of *Forensic Files*. Before the fizz had subsided in the drink I had just poured myself, I noted the familiarity of the Des Moines skyline on the TV screen. The storytelling voice of the narrator eased me into the program, and the flash of a date on the screen—August 23, 1993—flooded my senses.[1]

Despite the passage of nearly thirty years, I was drawn into crime scene photos, recalling the horrific murder of a thirty-six-year-old businesswoman named Patricia Lange, who had relocated to Des Moines from Colorado, eager to pursue a new position in the mortgage field. She had checked into a Holiday Inn on a busy thoroughfare in suburban Des Moines—a prime location with easy access to I-235 and I-80—for a week's stay to search for an apartment.

I had been in the hotel on numerous occasions. It resembled the layout of an Embassy Suites floor plan, the open center of the complex an opulent setting for brunches or professional events. Hotel rooms flanked the perimeter on all sides. It felt safe, as one could circumnavigate the hallways, all the while in the hearing distance of guests and hotel workers in the cavernous lobby.

Chapter 5

Sometime in the hours of Saturday evening or very early Sunday, Patricia Lange had been raped and strangled, her rigid corpse lying on the floor next to her bed. Robbery wasn't the motive, as two hundred dollars remained in her purse. The investigators theorized that she had been attacked by the perpetrator as he came out of hiding within her room or had gained her trust to open the door and allow him to enter.

The most promising lead in the case came from a thirteen-year-old girl named Candace Weiland. She had traveled from her small Iowa town, along with her mother, to compete in a talent show at the Iowa State Fair. As she entered an elevator on the ground floor of the hotel, she felt wary of the single male occupant beside her. He leaned in and was watching to see which floor button she pushed. When she exited the elevator, she turned back to see he was following her too closely. Candace's intuition kicked in, and she broke into a run, yelling, "Hey, Mom! I got the magazine you wanted," as she reached her hotel room door.

I was impressed with the young girl. She had done everything right—she was aware of her surroundings. She ran from a potential predator and signaled very loudly that she would not be alone when she entered her hotel room.

As the next morning unfolded, it became clear that Candace may have interacted with a prime suspect in the case. Until recently, Donald Piper had been the chief engineer and head of maintenance at the Holiday Inn. He had been fired weeks earlier on the grounds that he made unwanted advances on the women who worked there, and a sexual harassment suit had been filed against him. Donald Piper was heavily questioned, but ultimately, he was not brought up on any charges.

The multisensory opening to the show—the recognition of footage of my hometown and the flash of the date on the screen—opened my mind to connections I had not previously made. As the hairs on my arm rose in shock and the ice in my glass clinked in my trembling hands, I realized I knew precisely where I was in the week preceding this horrific homicide. A man bearing a striking resemblance to the man suspected in the Lange murder had approached me—a traumatizing experience I could recall nearly thirty years later. After much thought, I opened my laptop and sent an email to the detective profiled in the *Forensic Files* episode—Agent John Quinn, now the police chief of Waukee, Iowa.

Dear Police Chief Quinn,

My name is Jill Viles, and I happened upon an episode of Forensic Files, *"The Murder of Patricia Lange." The show prompted me to recall an unusual event that happened to me in the week leading up to Ms. Lange's murder. At the time, I was eighteen years old and participating in rush at Drake University in the hopes of joining a sorority. We walked in groups as we went from house to house, but at one point, I was separated from my group. It was at this point, I encountered a man I believe may have been Donald Piper. I was confused and frightened by what occurred, but I didn't tell my parents what had happened; however, I did talk about the incident with my sister several years later. When the date flashed on the screen, I was shocked by the memory because I could place what happened to me in relation to the tragic events that happened at the Holiday Inn. I hadn't previously connected the dots.*

I can offer greater detail, but I don't know if this is relevant in a case this old and one in which the person in question is incarcerated. However, I sense this was a very important case to you personally, as well as professionally, and I wanted to reach out to see if you wanted more information.

To offer more information about me, I have been married seventeen years, and I am the mother of a sixteen-year-old son. I work as a substitute teacher for the Southeast Valley School District. I am currently writing a memoir. I had lived in Des Moines all my life until moving from the area when I married in 2005. I am currently forty-seven years old. My father is Robert Dopf. He served many years as an assistant U.S. attorney for the Southern District of Iowa.

Sincerely,
Jill Viles

1982

The status quo had existed for decades in Iowa, with the blinking of the awakening streetlights at dusk indicating it was time for the children to return home. In the hours between the end of the school day and the evening meal, school-age children were free to roam—climbing chain-link fences in a game of hide-and-seek or assembling a pickup game of softball on a dead-end street—that is, until the early-morning hours of September 5, 1982, when the innocence of suburban Des Moines was shattered by the abduction of a paper boy named Johnny Gosch.[2]

Johnny Gosch typically delivered his papers with his father, but on this occasion, he slipped out for his route with the family's miniature dachshund, Gretchen. Just past 7:00 a.m., an irritated customer called Johnny's house to inquire about his missing paper, followed by a forlorn Gretchen returning alone, leash trailing behind her. Eyewitness testimony described two males questioning a twelve-year-old boy for directions and then shoving him into a vehicle. An abduction by a stranger was something almost unheard of in this affluent community, and no one gave a second thought to there being any danger to a preteen delivering morning papers.

The tragedy was a flash point not only for Iowa but for the entire nation as the cases of Johnny Gosch and thirteen-year-old Eugene Martin, who was abducted in nearly the same manner a year after Gosch, created panic for parents. The local dairy, Anderson-Erickson, printed photographs of the two young boys on cartons of milk, and numerous dairies across the country followed suit.

"Blue Star Homes" were introduced, those in which the home environment was declared wholesome and a responsible parent would be available in the hours after school. Those homes certified as "safe" were given a placard with a blue star on a white background to hang in a front window.

My experience with the "stranger danger" crusade that soon followed took place mainly within the frenzied assembly in my elementary school auditorium. With the enthusiasm of a championship referee, a police officer leapt about the stage, asking what we should do if we encountered an adult we didn't know, to which my friends, sardined with me in the wooden pew, would leap to their feet and scream, "Run!" as they vigorously jogged in place. I took my cue to slip beneath the seats, wondering what I was supposed to do since I couldn't run. The tanned athletic legs of my peers pumped beside me as I pondered my chameleonlike limbs—large bulbous veins rising to the surface of my limbs in search of absent flesh, like the threadbare skin of the elderly. I had a very large problem to solve.

I found my voice in the cacophony of a passing train—the steel-grating, land locked–rocking, serpent-hissing specter, winding through the darkness of a humid Iowa night. It was then that I locked hands with my dad, putting aside my self-consciousness, and belted out my scream with operatic flair. I had the power within me, he explained; I just had to set it free. Afterward, I was treated to an ice-cold strawberry-flavored Crush soda pop, the fizzing bubbles quenching my vocal cords like bouquets of flowers tossed onstage for a brilliant vocal soloist.

Away from the train, my screams were far less fruitful. My dad wanted the frenzied cat-tail-beneath-a-rocking-chair yowl, not the cotton-candy "eek" that escaped my lips, most befitting of his pet name for me, Pip-Squeak. Try as we might, the slosh of the whirling dervishes at the car wash, a tandem ride down a snow-covered hill, or the brief milliseconds we were airborne on the giant slide at the Iowa State Fair couldn't touch the loss of inhibition I experienced with a passing train.

Perhaps it was the proximity of unbridled force that unlocked something within me—the gentle rocking of our vehicle, the slender crossing bars unfit to hold back the 70-horse-powered engine of our car, the knowledge that a slip off the brake pedal would launch us into the train's path. It was a reminder that danger was omnipresent and unpredictable. It would appear out of nowhere at the most unexpected of times.

We screamed for greater reasons than the thrill it gave me deep in the pit of my stomach. There was a method to this madness—a lesson my father hoped would carry me through the ages, even at the points in my life when he wasn't there to save me.

He had good reason to fear for my safety. While other children could run from danger or buckle an adversary with a swift kick to the shin or jab to the nose, I had no such defenses. He'd thought ahead, imagined the worst, and considered my options when cornered by danger. I had just one weapon in my arsenal—my voice—and he needed me to fine-tune this asset, wield it like a weapon, and banish my inhibitions to the far corners of my mind.

To free me from my inhibitions was no small task. I took to rules and schedules like water to a sponge. I was the three-year-old child who would separate each toe, one from the other, wiping away the most minute speck of dirt or piece of lint before climbing into bed; the first grader paralyzed in my tracks in the school hallway as I realized I'd entered school chewing gum; and the grandchild who refused to sleep without my undies no matter how much my grandmother protested.

1984

My dad almost never worked from home. His office was tucked into a corner of our basement, sharing space with the "art room," a collection of watercolor paintings pinned to two rows of clothesline. Occasionally, I'd tiptoe into his sacred space, curl up in his office chair, and explore the contents of his office desk. In one drawer, I found a box of cigars. I shook it, and the cigars rolled about like stacked lumber, a testament to his sparse use of tobacco. In another drawer, I found a wallet containing a photo of a newborn me in the hospital nursery and a 1975 pilot's license—their compartments, made of virgin leather, secure from the fading of sunlight and the smudges of handling. My favorite pastime was to recline in his chair, with a cellophaned cigar in my mouth, and kick off from his desk, spiraling the office chair on the plastic floor mat as I imagined what he thought about in this private enclave of our home.

I remember my dad joining us for supper and then heading to the Drake University Law Library to work on an upcoming prosecution. There, beneath the glow of a green-shaded library lamp, he would sift through crime scene photos, interrogations, and witness testimony, piecing together a narrative that would secure a federal prosecution.

In the mid-eighties, he prosecuted the most consequential case of his law career.[3] The details of the rampage were heartbreaking. A family in Ottumwa, Iowa, was selling a used vehicle. On the pretense of checking out the vehicle for sale, two men—a drifter in his early twenties and his uncle—entered the property. In a horrific turn of events, the men broke into the family's trailer; tied up the parents; and raped the teenage daughter, who was six months pregnant. The mother was stabbed in the neck, and their daughter, Kathy Allen, only twelve years old, and their pregnant daughter were forced to accompany the fugitives on the run. In a desperate bid to save her life and the life of her baby, the teenage girl leapt from the moving car.

Shortly after crossing the Iowa border into Missouri, Kathy's throat was cut, and she bled out in a roadside ditch. Her last word to her attackers was that she wanted them to promise she'd be home in time for the Special Olympics. The decision to cross state lines sealed the fate of the two men. The miles they traveled into Missouri launched the case as a federal crime—my dad's first death penalty case.

Years later, on an afternoon as I sifted through VHS tapes looking for a show I'd taped, I came upon an unlabeled cassette. As I fed the tape into the VCR, I was alarmed to see footage of a crime scene—tall, blood-covered prairie grasses waving in the breeze and dozens of FBI agents securing yellow tape around the ditch. Fortunately, I tugged the tape out in the nick of time—I never saw the victim, but I will never forget the evil of the scene.

SUMMER 1992

To save steps and curious questions from passersby regarding my unusual gait, I maneuvered a well-worn path, suitable for the passage of a single pedestrian, as it snaked through a vacant lot adjacent to the Drake University campus. I'd been accepted to take courses early, though I had not yet graduated from high school. I loved the independence I felt as I explored the campus, but with a sudden fall, I realized my predicament—seated at ground level, not a speck of me could be seen, as the tall weeds camouflaged my body. Slowly, I ascended from the ground, moving palm over palm up my knees and hips to generate the leverage and force to stand.

As I pushed off from my thighs with a violent thrust, I knew this would be the last time I would rise independently from the ground. I'd passed these milestones over the years—the last time I'd step onto an escalator, the last time I'd step off a curb, the last time I'd attempted to raise a baby to my shoulder—occasions on which I made the mark but just barely. These conciliatory adaptations were a sharp contrast to my life circumstances. I was seventeen years old, a time when the last thing on my mind should have been a symbolic gauge that blinked to signal that I was nearly out of gas.

With these changes in mobility, my dad resurrected his safety surveillance. He often drove me to and from my part-time work as a hostess at the Drake Diner, a fifties-style diner buzzing with excitement and neon lights, a half block from the Drake University campus. Stepping out of the diner's warm glow into the small parking lot where my dad often waited, reading a book in his low-seated red sports car, I was torn between appreciating the love my dad had for me and the wish to gain my independence.

"Really, nothing is going to happen here," I'd lamented as I sunk low into the front seat and the automatic seat belt crossed over my shoulder.

Weeks later, a clip of the diner surrounded by a swarm of police car lights flashed on the evening news. Though I'd discontinued my employment as a hostess with the start of a new semester, the scene felt all too real. I took in the horror of the crime—two managers shot to death at the hostess stand, pools of blood where I'd stood just weeks ago.

I imagined myself in their place—cheeseburger, fries, milkshake, cash tendered—my shaking hands manipulating the cash register—a transaction entry required to spring the cash drawer. I would never have made it in time, though I have no doubt the managers would have pushed me to the floor as they had junior staff to save them on this fateful night.

Without words, my eyes conveyed the lesson I'd been taught on that night. The robbery suspects were apprehended days later partying in a downtown Des Moines hotel. The getaway driver, a cousin of the robber, was a kindergarten classmate of mine.

As I participated in sorority rush at Drake University, I wanted more than anything to pass this litmus test of normalcy. On my first day, the all-important house tours—a day my physical resources would be tested to the maximum—I was delayed by running errands with my mother. As I realized we were nearing the campus with minutes to spare, I burst into tears.

"I can't add one more thing, Mom," I cried. "I can't be late, not on top of everything else."

"This is ridiculous," my mom countered. "You have just as much chance as anyone else. This is all in your head."

I knew my mom meant well. This was the same narrative we'd bounced back and forth throughout childhood. She'd emphasized the positive spin, hoping that an enthusiastic vote of confidence would lend me the chance to shake off my fears. I couldn't help but notice the shifting of eyes of someone meeting me for the first time and the lulls in conversation as I navigated a staircase with a two-fisted grip and numerous stops to catch my breath.

As I blinked back tears, I joined my queue gathered at our meeting spot. I did my best to mask my exhaustion as I climbed numerous staircases and balanced decorative treats and cups of punch as I engaged in friendly conversation.

As we exited a home off the traditional Greek grid, my legs quaked with exhaustion, and I fell behind my pledge group, waving my fellow pledges onward because I didn't want to slow them down.

In the distance, I could hear the overly bright clapping and serenades of the various Greek houses. I pressed on. As I closed in at fifty meters up the street, I paused. Masked by the foliage, I could watch the welcomes but avoid being noticed for my tardiness.

Out of the corner of my eye, I saw movement, then a body closing in—too fast, too close—too fast—too close—my heart drummed like a hummingbird's wings. In an instant, a man nearly ran through me as if I were vapor and he could extinguish me. Without a word, he pulled his penis from his jeans and began masturbating vigorously an inch from me. I was paralyzed with horror having never seen anything like this.

I wondered why this assailant hadn't veered toward the dozens of girls down the block. Wouldn't this creep want the maximum reaction of screams? And then it hit me. My attack wasn't random. Most certainly, he had been watching me—maybe for seconds, maybe for several minutes. I had been targeted because I was weak. I had assumed the plight of the injured gazelle, the one separated from the herd with a lame leg, the one you root for in a documentary—"Run, Bambi, run!" though you fully know fate's been decided long ago in the reels of footage lying on the cutting room floor.

Any normal eighteen-year-old would bolt for safety, but I remained glued in place, the shame of my predicament filling every cell of my being. I was trapped alongside a simple street curb, something I couldn't climb, no matter my desperate need to get away. Could anyone see me? Did this beast of a man have a car idling to push me into? Could he shove me into a bush?

As the sorority members and potential pledges sang the last verses of their welcome songs, they paired off two at a time to enter the houses. In moments, the street would clear out, and I would be alone. With a wicked intake of air, I screamed as I had never screamed before, startling the coeds from song four houses deep into Greek street. The street grew eerily silent; everyone turned in my direction. Three fraternity members abruptly paused their football practice and ran toward me, squinting in the first steps toward the darkened canopy of bushes and trees. As they came closer, the man let loose profanity with the angst of a bull led into a rodeo pen. The fraternity brothers gave chase, and a circle of Pi Beta Phi sisters enveloped me and led me into the house I would pledge days later.

"We heard you scream in the house!" one member explained as I attempted to catch my breath and calm my nerves.

Thirty years later, I considered my actions, and my inaction, in this terrible circumstance. Why hadn't I reported what had happened to the police? To my parents? Why did I fail to seek counseling? Did my physical disability make me a greater target of sexual violence? Most importantly, I wondered if there was anything I could have done to prevent the tragic events that unfolded days later.

A few weeks after I had reported the incident, I received the results of the investigation in my inbox:

Hi Jill,

Wow, what a traumatic experience for you at such a young age. I know how difficult sharing a memory like this can be. I want you to know that even if at the time the incident happened you had reported it to police, it wouldn't change a thing regarding the Pat Lange Homicide or any of the other women that he hurt.

Your Father Robert Dopf was a great man who positively impacted the lives of those individuals that met him. He was a very noble man that epitomized all that is good about the profession of law and law enforcement. My heart was heavy with sadness when he passed.

I wish you and your family the very best. And, if you have any questions, please don't hesitate to reach out to me.

I hope that you and your family have a blessed Holiday season.

John F. Quinn
Chief of Police
Waukee Police Department

My first semester living away from home was plagued with dreams about my dad. Each night, they started the same way. I found myself in a long hallway lined with uniform doors on each side, like a hotel corridor. At first, a low rumble echoed through the passage, but as I traveled forward, the sound became more distinct: "lub-dub, lub-dub, lub-dub." I recognized the familiar sound of a beating heart, but this heart pumped lethargically, winding down like a mechanical toy. I furiously pried at each knob, but none would give.

For reasons I can't explain, I sensed it was my dad behind one of the doors. We had had a few conversations about the irregular beating of our hearts, and I confided in him that sometimes I felt an odd sensation in my chest, like a fish out of water flopping about.

"I wouldn't worry about that," he replied. "My heart did the very same thing when I was your age."

I shared these dreams with friends at college over lunch.

"Maybe you're a bit homesick, and that's why you're having these dreams," one friend offered.

It seemed a reasonable thought, but a few weeks later, another odd dream awakened me. This time, I heard the whir of an engine and the screech of metal. An airplane crash? I awakened in terror.

To my relief, I uncovered the mystery of this phantom airline accident. As I drew my dormitory curtain aside, I witnessed the emptying of the dumpster into a garbage truck just outside my window. So much for my clairvoyant skills. I had an overactive imagination, that was all, and I resigned myself to stop reading too much into my dreams.

With the start of a new term in January, the "hallway with dozens of doors" dreams returned with a vengeance. This time, I'd awaken to my own heart beating out of control. Deep in my chest, there was a swift kick, like the beating of a bass drum, then the rhythm returned to normal. The experience set my sixth sense ablaze.

I scheduled an office hours appointment with my genetics professor, Mike Myszewski. As I rummaged through my sack lunch, filling my stomach with bites of bologna sandwich and an apple, I did my best to share my family's genetic pedigree as my professor jotted notes. At one point, he stood and surveyed the bookshelf on his wall with a look of perpetual contemplation as if one of the texts would leap toward him by osmosis, opening to a page of interest.

"We're so rare. We're not even in one of these books," I suggested.

Sheepishly, he returned to his office chair beside me.

"I just don't understand how this could happen," I began. "How does this happen to four out of five siblings?"

Professor Myszewski's face illuminated like a light bulb in a socket. This was a question he could answer. Deftly, he performed a series of calculations, then turned his legal pad to face me. The answer was one out of sixteen. This is how often one would expect to see four out of five children affected by a dominantly inherited condition, each conception, a fifty-fifty coin flip.

To his chagrin, his calculations opened a floodgate of tears.

"No, I don't mean *that*," I sputtered. "I mean, *why* did this happen?"

I wasn't floundering in the world of logic and equations; I was asking entirely different questions. Why is there suffering in the world? Why do bad things happen to good people? We could invite the most prolific theologians in the world; complete a deep dive into the Bible, the Quran, the Buddhist suttas; and we would be no closer to finding an answer.

"Something's wrong with my dad's heart," I said, for the first time voicing my fear.

I knew I was rapidly losing credibility, as I eagerly accepted the Kleenex box passed my way and opened up about my peculiar dreams.

"You've been seen for fifteen years at the Mayo Clinic," Professor Myszewski reminded me. "They wouldn't have possibly overlooked something as important as a serious heart defect."

My only memory of cardiac investigation was during a family visit to the Mayo Clinic in 1985. My dad had been fitted with a Holter monitor to analyze his heart function. To entertain us, he pretended he was a Ghostbuster and that the blinking lights on the unit were part of his proton pack. This worked magically for our young minds and kept us free from worry. My dad had a fondness for shouldering the troubles we would face.

In the four and a half minutes it takes to make microwave popcorn, my family's destiny took an ominous turn. Purposefully, I'd left the journal articles inside my biochemistry text and stacked several other books on top as I stepped out of the room for the commercial break. As I emerged from the kitchen, shaking the bag at arm's length, releasing the warm, buttery scent, I came upon my dad engrossed in one of the articles.

"Don't look at those," I said hastily, but it was apparent I was too late. Protectively, my dad pinned the article to the table with his hand.

"I have all these symptoms," he said. "Bradycardia and AV block . . . they told me this was due to a virus."

I could never again play the card of ignorance, and neither could my dad. Up until this point, I could juxtapose any number of reasons to dismiss what I'd uncovered in the university library—I was too young to know better than my doctors, I had no formal medical training, the Mayo Clinic had followed our family for years. Surely, they would have identified a cardiac flaw and thereafter, the Muscular Dystrophy Association's Care Clinic—these physicians, nurses, and therapists focused exclusively on neuromuscular disorders. How could they possibly miss a diagnosis as obvious as Emery-Dreifuss muscular dystrophy when there were only nine subtypes of muscular dystrophy from which to select?

It was difficult to take the first breath following dreadful news. There was no one there to guide us—no patient pamphlet, no doctor's notes on a whiteboard, no after-visit summary or follow-up plan. The medical journal pages fanned out before us like tarot cards we hadn't wanted to see but, nonetheless, burrowed deeply into our shared subconsciousness.

I could read my dad's thoughts in the twitch of his jaw and the myriad contemplations on his face. He wasn't just stewing in this sobering news for himself; he was contemplating the lives of the four children he had fathered who shared the same genetic fate. No parent should have to take in news this way. It should have been relayed to my parents in a quiet, child-free moment back in our initial genetics consultation in 1979. Then I was a four-year-old child in the care of a cheerful nurse, a smiling face meeting me low to the ground as I sifted through a box of Saf-T-Pops, selecting all the red ones for me.

6

UN-WALKING

I think about this person living far away, yet in a strange way, inhabiting the same body as mine. I wonder what she does for a living, whom she loves, and what her family life is like, but, most of all, I long to meet her someday. I imagine traveling to a small café in a distant country and out of the crowd of bustling patrons, eyeing someone with the same waddling gait as me. Oblivious to the passage of time, I would speak endlessly to her like a woman reunited with a child given in adoption.

—Jill (Dopf) Viles, "Witches' Fingers Grab My Legs,"
Johns Hopkins Magazine, November 2000

I sent my message in a bottle with little hope of a reply. I had wanted more than anything to meet someone like me, but this was a dream I never imagined would come true. Then one evening in the summer of 2001, my world changed forever.

"It's me, it's Mary," said the cheery voice on the other end of the phone.

I listened intently as Mary described her symptoms of Emery-Dreifuss muscular dystrophy. For the most part, we are a nearly identical match. We share the same stiffened upper torso, the frozen elbow joints locked in a flexed position, and the curious waddle, and we've both experienced muscle weakness since early childhood.

"I thought I was the only one," suggested Mary. "All my life I was told I was the only woman in the US with this disease."

"Me too," I agreed.

"I never thought I'd live long enough to talk to someone who looks like me," Mary explained.

Within minutes, we laughed with one another with the ease of two friends who have shared a lifetime of memories. We talked for nearly an hour that first night, neither of us wanting to end the conversation. For the first time in my life, I no longer felt alone. We made plans for me to make the drive from Des Moines, Iowa, to Marseilles, Illinois, so we could see one another in person.

I knew I was being watched and scrutinized, but on this day, things were different. For the first time in my life, I don't hear the furious scribblings of ink pens as I walk for a group of doctors. Behind me, I heard the hum of conversation and the whir of a motorized scooter. As I turned to face forward, I saw the warm, friendly smiles of friends I felt like I had known for a lifetime.

"That's the walk, Mary," purred Mary's husband, Terry, as he stroked his chin. "That's just how you looked when you walked down the aisle."

Mary and I studied one another's movements closely. We are mirror images of one another—me, an image of Mary twelve years earlier, and she, a reflection of my life twelve years into the future. At ages twenty-seven and thirty-nine, respectively, our bodies have been changed in ways different from most others', yet this is the connection that has drawn us together. This is the first time either of us has gazed upon another person with Emery-Dreifuss muscular dystrophy.

Within minutes, the large garage owned by Mary's in-laws began to fill with family and friends. I tired as I stood, and instinctively, I slid my palms into the back pockets of my jeans and locked my thumbs in the belt loops. This subtle movement in posture shifted the bulk of my body weight from my calf muscles to my hamstrings. Mary's friend Karlene shook her head as she noted the overly worn pockets and frayed belt loops of my jeans.

"Mary used to stand just like that," she said. Karlene had been looking out for her longer than I'd been alive. They had met in grade school in a physical education class.

From the far corner of the garage, the smell of seared pork and portabella mushrooms filled the space. Terry cooked in a makeshift kitchen in the corner of the garage. From among the tools on the wall and the cans of motor oil and spare screws and nails, he drew tiny canisters of spices. He seasoned the food and checked it for taste before presenting plates of food to his dinner guests.

I took a seat in an old, faded recliner alongside Mary's scooter. Her face beamed with excitement. Mary's long blonde hair was styled in thick curls that fell to the center of her back. Even as her muscle weakness progressed, I knew she spent hours each morning getting ready and lingered in the tub for long, relaxing bubble baths. I cherished the way she cares for her appearance even as her heart faded within her chest.

As Mary departed to get ready for bed, I remained behind in the garage talking with Karlene. She shared her favorite memories of Mary in between soft puffs of her cigarette and spoke with a fondness I've only seen expressed by a parent of a small child. Soon, Terry rejoined us, and I gladly accepted the offer to hold on to his arms as I crossed the garage.

"I know you want me to hold my arm high—just like this," he said as I grasped his forearm just below the elbow.

By our mid-twenties, both Mary and I had developed the tendency to grasp hold of a person's forearm when walking. It was done partly for balance and partly to ease the throb of weakened muscles. A cane or walker was positioned too low, but a forearm meets at just the right height.

As we reached the doorway of the garage, I stopped in place. Though a drop-off of only a few inches existed on the opposite side, I extended my right foot before me and tapped three times before I stepped through the threshold, like a person with a visual impairment tapping the floor with a cane. I gasped as I released my grip on the doorframe, noting that the oils from Mary's hands mirrored the placement of my own hands.

Later, as we approached Mary's house, Terry leaned against my car, folding his arms across his chest.

"I know what you're thinking about right now," said Terry. "You're worried about tripping on the sticks on the sidewalk. You are already plotting your path around the pothole in the drive. But most of all, I know you're eyeing those three steps and thinking about how you don't want to climb them."

"Exactly," I said as I laughed.

"But when you come for a visit, you don't have to worry about that anymore."

Terry secured my arm around his neck, and in a single motion, he picked me up from the ground. "Remember, I've done this only about a million times."

An unsteady walk that would have taken minutes passed in a matter of seconds. The ground spun beneath us, a haze of brown and green. Terry's steady steps crushed sticks and flattened grass. Inside, I peered over Terry's shoulder. The living room was dim and unoccupied—a pair of shoes rested idly in the corner, a jacket draped over a folding chair, the room's solitary piece of furniture.

The central corridor of the house was ablaze with light and sound. As Mary's ability to walk became impeded over the years, walls were removed and extra doors installed to secure the fewest steps to any destination. Over time, even the walls of the small white bungalow seemed to wrap around the spirit in its central bedroom.

Mary was already dressed for bed. She wore a long, silk nightgown and a cotton cardigan, and she sat like a Barbie doll upon the mattress. Her long legs stuck straight out before her, and she didn't recline into the pillows behind her. As Terry sat me on the bed beside her, I naturally assumed the same posture; it lessened the strain on my back muscles.

Kelsey, Mary's thirteen-year-old daughter, climbed across the bed with the eagerness of a young child seeking sanctuary from a thunderstorm. She burrowed her face in the crook of Mary's arm and spoke softly of the day's events at school. The ice clinked in Karlene's glass, and she stepped outside the bedroom doorframe before lighting a second cigarette. Kelsey's back faced me for the duration of their conversation.

"You're such a pretty girl," said Mary as she ran her fingers through Kelsey's silky blonde bob.

Their relationship was unlike any other mother–daughter relationship I had known before. When among others, many mothers would briskly brush aside their daughter's concerns and suggest a time in private to speak, but Kelsey and her mother have persevered against the backdrop of a bustling ER and ambulance sirens, securing moments of serenity amid chaos as Mary's transplanted heart weakened.

Mary had a heart transplant in 1988 at the age of twenty-four.[1] It was the third ever performed in Peoria, Illinois, and now, nearly fifteen years later, only three of the original thirty-three patients with transplants in 1988 were still surviving. Though doctors cautioned that she would most likely survive only five years, she surprised even herself when she reached her tenth wedding anniversary in 1997. Mary had wanted, more than anything, to meet the family of her heart donor, but the records were sealed. She knew only that her heart came from a twenty-four-year-old woman who died in a motorcycle accident. When she read an article about me, she'd wanted to reach out to me in place of her heart donor.

"Girl, we have got to get you married," said Karlene as she took a long, slow drag on her cigarette and smiled in my direction. Karlene had been looking out for Mary longer than I've been alive. She knew what was in my future, and I trust her more than I trust my own doctors.

As I prepared for bed, Karlene passed before the open bathroom doorway and cupped her hand over her mouth in surprise. I was embarrassed that she had discovered me seated on the porcelain of Mary and Terry's room, which meant that Terry was sleeping on a sofa. Later, as I lay in bed beside Mary, I realized what it must feel like to sleep beside me at night. Due to our weakened muscles, we both overexaggerated our rolls to one side or the other. Mary thrashed about like a fish upon shore, and I knew that I moved in the same manner.

"I'm so cold," Mary lamented moments later.

Mary began to snore softly, and I realized she had spoken out in deep slumber. To my chagrin, I realized I had curled into a ball at the far corner of the mattress, taking the covers with me. As I draped the sheets and blankets across her body, my hand grazed her leg, and I felt a chill, like a palm grasping a metal doorknob on a frozen January morning. She had warned me about the coldness of her limbs, a consequence of heart failure, but I hadn't understood the reality until I brushed against her. Strained gasps of air escaped her lips, and her breath was labored and guttural. I lay close to Mary, with our intertwined hair pooled into a wreath above our heads as our chests rose and fell while we breathed in unison.

"I love you, Mary," I whispered.

"Good morning, Annabelle."

Sunlight streamed through the curtains. The shadow of Terry climbed the bedroom wall as he knelt beside the bed. In the distance, I heard the soft splashing of water filling a tub. I laid motionless as if I were asleep. Inches away, a model of my future body shifted position on the mattress. I resisted the instinct to roll over, and I gauged Mary's strength as she woke beside me.

Mary's walking stick tapped with each step as she traversed the ten-foot journey from her bedside to the adjoining bathroom. She entered the tub, and I heard Kelsey's cheery voice as Mary picked out bath gels and soaps from the dozens of bottles that circled the tub.

Terry left for work, and Mary continued her bath. I got dressed and laid back on the bed, tempted to fall asleep again, as an hour and twenty minutes passed. Perhaps, later this morning, a teacher might admonish Kelsey for a forgotten algebra worksheet. I couldn't help but realize that Kelsey had contemplated more about life and loss in these brief morning hours than most people ponder by middle age.

I heard the soft taps of her walking stick as Mary emerged unassisted from the bathroom. I longed to know how she walked, how I would someday walk, but I remained on the bed and watched Mary's movements through a crack in the bedroom door. Mary was worried that I would be upset to learn how she must move, to learn how the increasing weakness had affected her, and how it might affect me in the future. Instead, I felt a sense of relief as I gauged how she moved about her house. She could bathe and dress independently at the age of thirty-nine, and this was a tremendous reassurance to me.

"So where does the name Annabelle come from?" I asked Mary as I joined her on a stool beside her at the kitchen island.

I listened with great interest to the remarkable turn of events in her mother's life that had led to Mary's birth. She had been traveling with a friend, taking in the racing circuit in the South, and developed a brief love affair with a race car driver. When she told him she was pregnant, he'd offered to marry her, but she insisted that the racing circuit was no place to raise a child and that she needed to return to her hometown of Marseilles, Illinois.

Mary's mother managed to make some of the distance toward home, but upon entering Tennessee, she realized she was already in labor two months early. She was admitted to a Murfreesboro, Tennessee, hospital and prepared to give birth far from family and friends. Two nurses, Vera and Annabelle, took care of her, and she searched their eyes for strength between birthing pains. Moments later, two baby girls entered the world—Eileen Vera and Mary Annabelle—and so they were forever named.

Mary's version of her life story was far more interesting than the no-nonsense research article I'd come to know her by, an article Mary had never seen. As I showed the article to her, I was curious about her reaction. She furrowed her brow and bit her lower lip as she viewed a microscopic image of her muscle biopsy. Arrows pointed to areas of muscle fiber death and other disease states.

"Huh," she said in a disappointed tone, "so there's no photos of me and Terry? Nothing that describes who I am?"

"I'm sorry," I offered. "I guess it's just the facts."

Mary pondered the family tree outlined in the article, noting that some of the squares and circles were darkened to note the inheritance of a disease allele. A Chicago neurologist was so taken by her story that he drove the forty-five minutes to Marseilles, a small town southwest of Chicago, to examine Mary's seven half siblings as well. The physician asked Mary's healthy half siblings to remove their shoes and socks and stand; he noted that several had tightened Achilles tendons that couldn't extend the heel to the floor. He described this as variable penetrance—the same genetic variant that robbed Mary of her walking ability and her heart had barely touched several of her healthy half siblings.

"So where did this disease come from?" she asked me.

"If the gene were passed on to both you and some of your half siblings, that would mean that your mother also had this genetic variant," I explained.

Mary's eyes widened, and she swallowed hard, pondering the overwhelming facts.

"My mother who was strong enough to have nine children? You're saying this came from her?"

She looked intently out the window for a few moments with an expression both sad and perplexed.

"You know my mother used to get lost in her thoughts sometimes," said Mary. "I'd come up alongside the car, and I could tell she'd been watching me play with kids at school. She'd get a far-off look in her eyes, and she'd say, 'I worry about ya, Mary.'"

Mary shifted momentarily on her stool and brushed a strand of hair from her eyes.

"About once a year, she'd take me to the Shriners clinic. She didn't want me to have too many memories of doctors, so one day when we left, she offered to take me to Long John Silver's. I just remember her sifting through her purse and stacking coins on the hood of her car, then saying something about not being hungry herself. I knew we didn't have the money, but then she went to open the car door and just burst into tears—she'd locked us out. I can never forget seeing her crying like that. Sometimes I think she took it harder than I did, me having weak muscles."

She extended her walking stick in my direction. I shook my head.

"I don't want it," I said.

I was fearful of Mary's dependence on her walking stick; I knew she could not take a single step without it.

Seven years ago, Mary found that the kitchen countertops no longer provided sufficient support for her, so she'd started using her broom handle. With the broom handle, she's learned to take slow, steady pivots as if braced against a pogo stick that never left the ground. Returning from work one day, Terry watched Mary moving about the kitchen, broomstick in hand. That gave him the idea to craft a walking stick for Mary, and she's used it ever since. Over time, the harsh lines of the chiseled wood were softened by Mary's touch. This simple stick represented their love for one another, the blending of their worlds. In the years that followed, Kelsey would stand with outstretched arms, beckoning her mother to take just a single step without the stick, but she never did.

As I attempted to rest the stick against the countertop, Mary cupped my hands about it and stared intently into my eyes.

"I need to see you try."

Grasping the stick haphazardly, I positioned it far from my body, as a shepherd wields a staff. My outstretched arm quivered as my sweaty palm slid over the smooth chiseled wood. I gave her a bewildered look.

"Like this," insisted Mary as she prompted me to weave my fingers together.

I held the stick close to my body and took in the tender scent of Mary's lotions and bath salts. I wrapped my interlocked fingers around the smooth round stick, and they settled into the shallow grooves, still warm from Mary's grasp. As I stepped forward, the soft taps began to emanate from me.

"When will I need this?" I asked.

"Thirty-two. Things will change when you're thirty-two."

Mary is a representation of what lies in my future. As we compared stories, I realized our bodies have weakened at nearly the same rate. We both worked at fast-food restaurants at the age of fourteen, served as cashiers in our late teen years, and lost the ability to rise unassisted from the ground sometime around the age of eighteen.

Mary adopted the radiant smile of a new parent. There was still much to learn and to overcome, but she knew the promise of the future lies in these tedious steps. The moments of my first steps are lost to the innocence of youth, but today, we celebrated a gain far more precious. The legacy of her survival has been passed on—I have learned to un-walk.

I noticed that the cupboard below the kitchen sink was slightly ajar, a pile of thick paperback books, several bottles of water, and a box of crackers, visible. Mary kept the supplies handy in case she fell on the floor. She could not get up without assistance, so she kept items to busy herself if she needed to wait for Terry or a friend to help her up.

I wondered what I would keep in a spare kitchen cupboard when the time came that I couldn't get up from the ground without help. It reminded me of the game in which people name the three items they can't live without if stranded on the moon.

"You understand why Kelsey's graduation is so important to me," suggested Mary.

I recalled lengthy phone conversations about upswept hairstyles and silk gowns, dyed-to-match pumps, and beaded handbags. There was urgency in Mary's voice. This represented the goodbye she could plan, the one she could rehearse over and over in her mind.

"I know I won't see Kelsey graduate from high school."

"Don't talk like that, Mary," I insisted as I tugged my barstool closer to her. "You're going to get a new heart, and everything will be—"

I paused in midsentence. Mary's face represented the purity of freshly fallen snow. Tiny creases lined the corners of her eyes and lips like a settling twig in a pristine landscape. She was both stoic and serene.

"I have my heart," she said as she stroked her transplant scar just visible above her V-neck T-shirt. In this moment, I felt small and inconsiderate. I was taking the easy way out.

"Would you believe I still can't walk in front of Terry without putting on a robe over my nightgown?" said Mary as she smiled to herself. "He says, 'Mary, we've been married almost fifteen years; I know what you look like,' but I just can't do it, even after all this time. When I was about ten years old, a doctor had me walk up and down the clinic hallway. People were passing by, in and out of rooms, and I didn't have anything on but my underpants. I was just starting to develop . . . and I just . . . I just never forgot this. Do you know what I mean?"

My eyes teared up, and I swallowed hard. We were only twenty-seven and thirty-nine, but I felt like we'd experienced a lifetime of synchronicity.

"Just promise me, you'll never settle for anyone who would hurt you. You find a man as good as Terry, or you don't marry," she insisted. "I don't care how lonely you are. . . . I love Terry with all my heart, but if I didn't have this kind of love, Karlene would take care of me. All my friends would. You just wait. Promise me this."

As I gazed out the kitchen window, the sparkle of Christmas ornaments caught my eye. My recent Christmas gift to Mary danced among them—a wooden spool, painted to resemble a snowman. I had imagined she would store the ornament out of sight, taking it out only once a year to trim a Christmas tree, but I came to know its omnipresence. Mary's life was a simultaneous celebration of the seasons.

The front door squeaked, and I heard Terry's footsteps as he entered the front foyer. Mary's expression changed suddenly. She clasped her hand over mine, her voice hushed yet urgent.

"Never speak of this in front of Terry," she insisted. "Promise me."

I blinked hard three times. Terry's footsteps stopped behind my kitchen stool. Mary's golden curls gleamed as sunshine flooded through the kitchen window. Any hint of worry escaped her beautiful face. Behind her, a photo of a sun-kissed Mary and Terry smiling on the deck of cruise ship drew my attention. It partially covered the invoice from her recent Life Flight helicopter ride.

Moments later, we paused for photos on Mary's backyard deck and then circled the house to my parked car. While I was seated in the driver's seat, Mary leaned in to hug me. As I pulled away, I saw Mary's friends and family in the rearview mirror waving goodbye as I drove away.

At a safe distance, I let the tears flow, and I turned down a side street and pulled to the curb. I was afraid I would never see Mary again. As I leaned forward, my forehead resting on the steering wheel, I found a moment of clarity much like Marty McFly in *Back to the Future*—*I have all the time in the world! I can simply circle back and announce a change of plans.*

I pulled the rearview mirror to me and saw in its reflection that my tear-stained face betrayed my true feelings. I couldn't hide the fact that I was fearful of Mary's pending need for a second heart transplant. It was good that I left the visit full of smiles and warm memories. It was time for me to head home.

I drove a mere forty-five miles before stopping at a Pizza Hut in Princeton, Illinois. The waitress noticed my loneliness as I stared out the window and mindlessly played with the Parmesan cheese and pepper flake shakers.

"You've left someone you love," she said as she placed my bill on the table.

I gave her a wan smile and agreed it was the truth.

My phone rang on a Saturday afternoon, twelve weeks after my visit to Mary's. The call interrupted my lunch preparation, and I wiped my hands on a towel before picking up the receiver. The aroma of onion lingered, stinging my eyes.

"Honey, it's Karlene."

"How's Mary?" I asked in an overly cheery voice.

"Mary passed away last night."

I listened numbly as Karlene described Mary's passing during her second heart transplant. I knew she topped the list of the most critical heart failure patients in her region, but a heart had been found so quickly that I hadn't even known she had entered surgery. Each update had been positive, but as Mary was disconnected from ECMO, a machine that provides cardiac and respiratory function during transplantation, she went into cardiac arrest.

"What was the last thing she said?" I asked.

Karlene paused for a few moments, and when she spoke, her voice was shaky.

"She was alone with Terry. She just wanted him to promise her she'd be OK."

"He told her the truth," I said. "People enter heaven in different ways. Some people are abandoned; others are abused. He carried Mary there."

As my mother and I parked alongside the funeral home, I immediately sensed the significance of the event. Young people gathered on the front porch, and many others spilled out onto the lawn. I stared at the ground as I walked, trying to squelch the burning tears in the corners of my eyes. Before I reached the front door, I felt Karlene's warm hug. She offered me her arm, and I walked with one hand in Karlene's and the other in my mother's.

As I entered the foyer, I was comforted by a large easel displaying photos of Mary's sixth-grade graduation class. Standing shoulder to shoulder, clad in seventies-style dresses, the young girls blended, one into the other. All of the girls, except one, sported platform shoes with chunky heels. Immediately, my eyes locked on one figure. I ran my finger over the photo, tracing the slender lines of a pair of black leather boots. They reached midcalf, revealing only a few inches of leg below the hemline.

"This is Mary," I said without hesitation.

Karlene nodded. "She was always upset that she was too skinny. She used to hide herself in those tall boots."

In our phone conversations, Mary had empathized with people of slight build. Until the age of twenty-four, her weight had remained at only ninety pounds despite her five-foot, seven-inch frame. Each time following her pregnancy and her heart transplant, Mary had gained thirty pounds. She appeared healthy and energetic, even in her later photos, though her heart was failing.

There was overwhelming love for Mary as we continued into the funeral home. More than four hundred people had already visited. As I joined the line to pay my respects to Mary's family, I noticed the gasps and the pointing of others as they stepped out of line to take in a better view of me. I wasn't offended in the least. I understood that many were Mary's relatives who had heard of me but had not yet seen me. A gracious woman extended her hand to me but decided instead to wrap her arms around me in a bear hug. With her lips pressed close to my ear, she whispered that she was Mary's oldest half sister. Her hand lingered as she stroked my hair.

Reluctantly, I walked toward Mary's casket. My mother and Karlene offered a gentle squeeze of the hand. As I looked inside, my turbulent emotions subsided. I did not see Mary there. The form in front of me appeared plastic. Out of the corner of my eye, I spied two more easels of photos celebrating Mary's life. I was drawn away from the casket and toward the photos.

"I saw a shooting star in the sky the night Mary died," said Terry. "I took it as a sign. I knew she was all right when I saw that."

Mary's scooter and walking stick sat idly in a corner of the room. Mary lay as any other person in a casket. There remained no hint of her muscle weakness. In this moment, I realized I would live and die just as Mary did. Thoughts of a miracle cure or medical advance that would significantly change my life disappeared from my psyche forever. As I returned home, I promised myself I would store research articles out of sight and search the web for research updates far less frequently. Instead, I would learn to take more pictures.

I discovered my favorite photo of Mary. Shortly after their relationship began, Mary was photographed leaning against Terry's motorcycle. In the midsummer heat, the sleeves of Mary's T-shirt are rolled to the shoulder revealing the once-hidden contours of her tiny arms. She appears confident, happy, and full of life. She cocks her head back slightly as Terry leans down to kiss her.

"This is when Mary became truly beautiful," I said to Karlene as I pointed to the photo.

7

CONTAINS PHOTO

Do Not Bend

The tour didn't sit well with me. Perhaps it was the ubiquitous white walls, the buzz of the fluorescent lights, or the five doors, each opening to a tiny room filled with an office chair, a TV/VCR combo, a box of Kleenex, and a small trash receptacle. Clients of the dating agency buzzed about like bees in a hive, exchanging watched VCR tapes for new ones. Though the office was immaculately clean, the business shared too close of an aura with a sperm donation facility. As the minutes ticked by, my hope of finding love diminished, as did my patience for the loud thwack! of stacks of VHS tapes falling like Jenga towers.

I'd promised myself I'd keep an open mind with this brave new world of media-guided dating, but I felt like a contestant on the *Love Connection*, waiting for the host, Chuck Woolery, to count down the audience's assessment of my suitability as a dating partner into a quantifiable percentage. Part of me was excited to enter a "high-traffic area" of the love persuasion, but the rest of me wanted to slowly back out of the forgettable strip mall business that had sprouted up seemingly overnight. It was difficult to give up on the dream of a Hallmark-style, cheesy, destined-in-the-stars meeting, one that didn't involve the aid of an introduction service.

The idea of sitting alone before a video camera, red record button blinking inches from my face, dictating who I was and what I was seeking in a potential partner onto a videocassette with a "Be Kind, Rewind," sticker, felt disingenuous. Besides, how did I know what would happen to this recording? Would it be shoplifted under a bulky sweatshirt? Perhaps added to someone's porn collection? And then there was my worst fear—would someone I know watch my video?

Looking back on the experience twenty-five years later, it's easy to forget that the dating process used to be much more cumbersome than the swipe right, swipe left of today's matchmaking. As Drew Barrymore lamented in the film *He's Just Not That Into You*, there were so many ways to experience a cold shoulder in the game of love—"I miss the days when you had one phone number and one answering machine, and that one answering machine had one cassette tape, and that one cassette either had a message from a guy or it didn't. And now you have to go around checking all these different portals to get rejected by seven different technologies. It's exhausting."

I made the leap toward a dating agency because a friend in my church group, Jordan, had gone before me. She had shelled out the hefty membership fee of a thousand dollars and endured nine subpar dating experiences. On her tenth and final date, she was so afraid of her match's erratic behavior that she'd considered jumping from a moving car. When she asked the owner for a prorated refund, she was offered another tenth date, this time at no additional charge. She scoffed mightily as she power-stepped toward the door, then turned around swiftly as if a pair of invisible cupids had turned her heels 180 degrees. On a whim, she accepted this on-the-house final match, and several months later, she joined her freebie date at the altar to exchange wedding vows.

It's stories like this that kept my hope alive. Just suppose Jordan hadn't changed her mind. Her destiny hinged on those five seconds of contemplation.

Still, I couldn't shake the feeling that this wasn't the typical order of life. Love, or at least falling in love, was a complex and mysterious blend of pheromones and serendipity. I knew nothing of the dance of love. For me, romance novels read like cookbooks—I could assemble the ingredients—candlelight, soft music, bite-size fruit on plastic holiday spears—but I knew nothing of the nuance of reading another person or the give-and-take of a relationship. In terms of my dating history, I was nearly a blank slate.

The problem with dating with a physical disability is that the "elephant in the room" can't be pushed to the back burner of conversation until it's socially acceptable to take a deep dive into the "dark web" of life experience. Maybe that's date nine or seventeen, maybe even date twenty-nine. Imagine trying to navigate a first date with the most personal part of your life tattooed on your forehead—"I've served time for a DUI" or "I'm at enhanced risk for the breast cancer gene, but I haven't taken the test yet." These are important conversations, just not comfortable topics for a first date.

In my earliest days in college, I experimented with alcohol as a social lubricant. Initially, this seemed to work miracles—my poor coordination was assumed to be a consequence of drunkenness, while the strobe lights of dance music masked my sloppy movements.

On one occasion, I found myself in line for the restroom at a fraternity party. The line snaked down a staircase, and I clung tightly to the banister for support. A frat guy stood at the top of the stairs, his jokes falling flat.

"Why don't you come and stand by me?" he asked somewhere in my general direction.

The girl beside me stepped out of line and took the stairs by twos.

"Not you," he rudely bellowed. "Her," he insisted as he pointed at me.

The unchosen girl flushed with embarrassment and slid back into the line. He was the type of guy I avoided at all costs, but under the influence of alcohol, I ascended the steps and accepted the invitation to snuggle closely beside him. Frat guy continued with his storytelling.

Sober, I never would have allowed his hand to probe my shoulder and arm, but I was too late to intervene. He stopped his rant abruptly and pushed the sleeve of my shirt past my shoulder to reveal my emaciated upper arm, the shape of my humerus bone clearly visible beneath my thin skin. He shoved me backward, my head striking the wall. As I took a step forward, my hand massaging the back of my head, I discovered the shallow dent in the wall where I'd just stood.

"Are you on drugs?" he blustered, beer-laden spittle striking my face. "Are you anorexic?"

I trembled in fear. My eyes pleaded with the other ladies in line.

"No," I stammered.

"What is this?" he shouted angrily as he thrust my arm inches from my eyes like a dog threatened with a chewed slipper.

My toothpick arms had sparked intrusive questions since the fourth grade. I could never forget the sensation of wriggling fingers looping about my sunken flesh just above my elbow and the wide-eyed disbelief when my classmates gauged the circumference of my ankles. Here, I was entering the first passages of adulthood, my vestigial arm drooping from my body like the dying petal of a flower. I thought I'd left these moments of mortification tucked away in the recesses of my mind.

The words of the owner of the dating agency sparked my memory: "We do have a discount program for clients who are . . ."

The owner appeared to slink a couple inches lower into her custom ergonomic chair. She twisted her wedding ring on her finger as her eyes darted around the room.

"Oh, how shall I say . . ."

My lips parted as I considered completing her sentence for her. Disabled? Affected? Nontraditional?

"Difficult to place," she announced with certainty.

Swiftly, she reached for her water. She drank greedily, the ice cubes sliding to the rim of the glass, creating a dam against her teeth and producing an annoying sucking sound. With her free hand, she raised her pointer finger in a gesture for more time.

I considered myself in the plight of a feral kitten, one banished beneath the fold of a cardboard box. There's an air of distaste from the pet shop owner and a swift rejection as the other, properly fluffed kittens are led into the showroom window, basking in the glow of oversize light bulbs and delighted faces pressed to the glass. The kitten's oddity is heavily nuanced—perhaps a case of mange, a disorienting calico, an off-putting purr—but I, like the kitten, know the feeling of skulking behind a darkened corner. A misfit. An outlier. A deviant.

"I can offer a trial match for one hundred dollars," she suggested. "You can try it out. If there's a connection . . . fantastic, but if not, I won't hold you to the terms of the membership."

I paused with my uncapped pen over my checkbook. One hundred dollars was a great deal of money to me but significantly less than a $1,000 commitment. I figured I'd always regret not knowing what would have happened if I had passed on the opportunity. I took a deep breath and filled out the check.

This would not be my first brush with a dating agency. A few years earlier, I had joined a novel dating agency that focused on matching people with physical disabilities with other people with disabilities or allies—those without disabilities who were open to meeting someone with a disability. The introductions focused on pen pal correspondence. To be safe, I rented a post office box.

I had imagined this would be the safest place, emotionally speaking, to check out the dating world. I was very afraid of getting hurt, and I figured meeting and possibly dating someone who shared a physical disability would level the playing field.

One letter I drew from my PO box carried the message "Contains photo—Do Not Bend." Someone had taken great care to deliver their introduction as neatly as possible. I opened the mail, and just as I'd expected, I was treated to a photo of a well-groomed man accompanied by a succinct and intelligent letter of introduction. His clothes were neatly pressed—a buttoned oxford and khakis—and he offered a pleasant smile. From the complexity of his wheelchair, I suspected he required assistance in daily living.

The gentleman had been born with osteogenesis imperfecta, a rare genetic disorder resulting in bones that are easily fractured. I was familiar with the condition based on Samuel L. Jackson's performance in the film *Unbreakable*. I couldn't help but ask God why it was fair that someone who put so much effort into a first impression wasn't born in a body shown the same care in design.

I felt a sinking in my chest. I couldn't lead this man on despite his photo and letter. The reality of two people, both with physical limitations, trying to date began to surface in my brain. Besides the physical reality, there was the geographic burden—we were hundreds of miles apart. Sadly, I tucked the letter back in the envelope and placed it in my bag. I didn't date this man, but the impression he made lasted many years. I sure hope he found love.

I heard the approach of my hundred-dollar date before I saw him. The offending truck circled the drive leading to my apartment entrance as I waited by the door. Momentarily, I noticed the muffler, dangling precariously low to the ground, as the source of the noise.

As I joined my date in the cab of his truck, we launched into a conversation of little substance. We shared nothing in common, and I had to shout to be heard. How the dating entrepreneur had envisioned a love connection between us was beyond me.

"I have some tools to pick up," the burly man suggested as we pulled into a parking lot frequented by pickup trucks and SUVs.

"Are you OK waiting here?" he asked.

My stomach growled in protest. I had been hoping for an early dinner and perhaps an end point for this date.

"Here, I'll leave you with some music," he said.

As I watched my date cross the parking lot, I took in the somber tones of the unusual music score. It took me a few moments to sort out the message, but finally I understood—"Your silence is your sin; the life that's saved was once your own; every life deserves a lifetime."

I hadn't realized the pro-life movement had a soundtrack.

Momentarily, I felt a thud, and the cab of the truck lowered a couple inches.

"I've got a couple more tools to load," my date said. "I'll be right back."

Like my friend Jordan, I considered making a hasty exit. What was he doing with all these tools anyway, and why pick them up on a date? Did he plan to build a secret room in his basement? Would I become known as the handyman's victim on a real-life crime podcast?

Two more loud thumps echoed from the tailgate.

"So did you like the music?" he asked as he joined me in the cab of his truck and turned over the engine.

"Well," I began, "It's a heavy topic for a first date."

He leaned in closer with a befuddled, "Huh?"

"I said, maybe we could talk about pets or hobbies," I yelled over the noise of the truck. "You know, lighter stuff like that."

My ears were ringing by the time we arrived at the restaurant for dinner. As my date negotiated the terms of his coupon with the waitress, I took stock of the good things in my life. I had a career I loved as a teaching assistant in freshman composition classes at Iowa State University; I was working toward my master's degree in creative writing; and I hoped someday to be a published author.

Later, as our conversation turned toward my date's explaining he'd decided to return to his ex-girlfriend, I didn't feel a single pang of sadness. It was one of the first times in my life that I had decided to adopt dark humor, and I masked this newfound interiority. I was mighty pleased with my life, and it was great to feel this way.

When I share that I've experienced weakness since toddlerhood, one of two questions is bound to surface: Have I lived a normal life? or Have I lived an abnormal life? Either question is maddening because it assumes a black-and-white answer when the reality is that the barometer of my life experience spins like a weather vane in a gusty wind. If pinned down for a reply, the answer is yes and yes. Ultimately, I am still trying to sort this out for myself.

I don't claim the winner's sash in the pageant of inspiration. I can't say that I've become a better person due to my disability. If anything, I've become self-absorbed and walled off from the feelings of others, particularly those I love the most. On one occasion, my husband reached out his hand to tug me up in bed. Without considering the effect of my words, I said, "I don't want to have to need you."

The sentiment stung us both like acid rain.

"I didn't mean . . ." I stammered as his face fell and I rewound the audiotape in my mind.

The question of "normal" and "abnormal" irks me because it skirts the omnipresent feeling within me: aloneness. Aloneness, a permanent state of being, is distinctly different than loneliness, a temporary emotion. Loneliness is a universal trait, or at least, I assume it has been experienced by nearly every person on the planet at some time in their life. Aloneness is loneliness cloaked in shame—a heavy pelt that burdens the shoulders and contracts the body's soul.

Each year, I eagerly awaited delivery of the JCPenney Christmas Wish List catalog. I had dog-eared my favorite page—pristine white roller skates with an embroidered Mickey and Minnie Mouse twirling in roller skates beneath a glitter ball. I asked for the same gift year after year, circling the treasured skates and snapping the page to the refrigerator door with a magnet.

In preparation for my custom roller skates, I'd begun to transform my look with a pair of denim jeans, to which I affixed a roller skate applique on the back pocket. I'd snatched the jeans from the box of hand-me-downs from my cousin Diana. I had hoped to receive the skates as a gift beneath the Christmas tree, but after asking for them for several years and not receiving them, I figured it was a long shot. The jeans had been well loved, and the shoelace on the applique was missing. So I took the shoelace from a cast-off Cabbage Patch Kid shoe and rethreaded the eyelets with it.

I dreamed of an alternate reality in which I strode into Skate Country, our local rink, long laces tied together and draped over my shoulder, my signature skates thumping against my body. Like fourth-grade royalty, I would bypass the rental counter—the rows of bleak skates with orange laces, stained and stiff like army surplus tents and smelling of Fritos chips.

My mother hadn't wanted to disappoint me, but she feared my fascination with roller skating. I'd want to wear them at home, and no doubt, she had nightmares of my careening down our drive into oncoming traffic or stumbling over cracks in the sidewalk. As a compromise, I suggested that if I did own the skates, I'd be willing to store them on a high closet shelf, retrieving them only for the occasional party at the rink.

One Christmas, on the prowl for Christmas gift reveals, I'd shaken a large box and sensed the rolling of wheels inside. My sister, Janet, had a similarly sized gift, and we giggled with excitement about our secret loot. Finally, we would have personalized skates.

Later, as we opened gifts, I tried to mask my disappointment—my sister and I had received "Safe-T-Skates," a clunky metal contraption intended to be strapped onto shoes. When we tried on the skates in our carpeted basement, the design of the skates aggravated my lower leg weakness, locking with each stride and thrusting my body to the ground. I hated them. Soon, they occupied the bottom of the toy chest. We never asked for skates again.

Despite the roller skate debacle, the class parties at Skate Country remained a favorite part of my childhood. I remember the shrieks of surprise that wafted over the stalls as we aimed to land on the toilet seat like a newly gigantic Alice in Wonderland. Out on the rink, the hum of the music pulsed through my skates as I screamed, "Ghostbusters!" with a chorus of my friends, and we linked our lace-gloved hands as we imagined ourselves "Material Girls."

The beat of Billy Joel's "Uptown Girl" lit a flame in my heart, while the song "Downtown" by Petula Clark struck a tone of dread. The "down" I could handle—bending my knees to sitting position, but the "town" posed a problem—I had no ability to rise to standing from the ground, much less in roller skates. I massaged my ego with a visit to the snack bar as I pushed down my differences with a Dixie cup of Pepsi and a matchbox serving of Lemonheads.

As the throng of skaters moved in unison, dropping and rising like a school of fish gliding on a shared ocean current, the pinball machine lured me away from the flock. It was set to standby mode, and I eyed the footprint of a phantom player as I uselessly flapped the flippers. In the corner of the scoreboard, a dimmed image pulsed into brightness with the increasing score—the horns of the demon took shape, as did the shackles binding the young woman's wrists and ankles. She lay helpless in the arms of her captor, her long blonde hair spilling about her captor's sinewy muscles. For the first time, an orb of heat pulsed between my legs. The experience was regrettable and shameful. In an alternate universe, I imagined my strong legs glided on the dulcet notes of Petula Clark's song, and I was linked arm in arm with my friends, far away from the pinball machine.

Abundant preparation went into meeting someone new for a first date. I had to arrive at least a half hour ahead of time to scope out the parking lot, tracing the path from my car to the front door, paying close attention to the slightest change in angle or the spiderweb splitting of concrete—either of which increased my risk of tripping and falling—so not in vogue for a twentysomething.

I had to be on the lookout for yellow parking lines. As only someone struggling to hold on to the ability to walk knows, yellow parking lines are slipperier than white lines. (It must have something to do with the phosphorescent dye in yellow paint altering the coefficient of friction.)

As I entered Friedrich's Coffee, our prearranged place of meeting, I gauged the heft of the door and scanned the stone floor inside for runaway coffee beans or overturned corners of floor rugs, either of which could create a loss of balance and an embarrassing fall to the ground.

Thankfully, on this spring day of 2001, I'd beat the after-work crowd, and a tall table with twin barstools was mine for the taking. My backside reclined against the stool, the plush seat bolstering my posture forward, and I casually braced my elbow on the clear round tabletop filled with aromatic beans. Satisfied, I took inventory of myself—the carefully crafted pose created the illusion I could stand without a grimace on my face. I was, perhaps, leggy, at least in the ratio of my legs to my torso. For a few moments, I could pass for normal, allowing me fair play in this game of first impressions.

I was searching for "JOMY23." We'd exchanged several friendly messages over AOL until our virtual conversation took a decidedly wrong turn. I'd mentioned my muscle disease and my unusual gait, and I was floored to see JOMY23's response: "Why don't you just tell people you got shot?" In horror, with mouth agape, I nearly burned a hole in my computer screen monitor with my glare. I had just begun to share more about my medical history, and this perceived insult had set me back. JOMY23 profusely apologized and suggested meeting up at a coffeehouse. After several volleys of yes and no, I finally agreed to meet JOMY23, whose real name was Joe Vadakkan, which he shared with me online before I would agree to meet him at the coffeehouse.

Minutes after my arrival, a young guy entered the coffeehouse. Due to our spat, he knew very little about me; I have blonde hair, and I was mad at him was all he had to go on. Sheepishly, his eyes moved from table to table. He was handsomely dressed in a pressed short-sleeve shirt and pants and wore a white baseball cap with the brim pulled low over his eyes. As he made his way in my direction, I couldn't help breaking into giggles.

"It's you!" Joe exclaimed.

We continued our introductions, and I explained I have a distaste for coffee, though I enjoy the surroundings.

In our first of many coffeehouse meet-ups, Joe brought me a glass bottle of orange soda—Joe's Soda, so appropriately named—and a lemon poppyseed muffin.

"Now, I've got to explain the gunshot reference," Joe began. "What I meant to say is that you can be a storyteller. Say, somebody asks you something and you're not wanting to go there, you can make the story be anything you want it to be."

I listened with interest wondering where this was going to go.

"So, instead of a disease, you can say you got shot taking down a robbery suspect in a bank. Now, you're the hero of the story. You can be who you want to be."

Joe's smile was infectious. Though we had met only moments ago, he had an uncanny way of ingratiating himself into my life story.

In the few minutes he had talked about his life, I knew far less about him than he knew about me. Joe immigrated to the United States from India ten years ago. He spoke of Bombay, now modern-day Mumbai. I wanted to turn the narrative back to him. I asked about his early childhood years, about his siblings, his parents.

"Oh, no, no—my father, he was a very dangerous criminal. We had to leave by cover of darkness," said Joe warily.

I attempted a couple follow-up questions, but Joe wanted to turn the conversation back to me. I talked about my internship at Johns Hopkins last summer as I explored my family's genetic disorder, which became the subject of my master's thesis.

"I spend more time in the library than I do in my classes," I admitted.

In a conversation I will loosely remember, Joe described the tradition of arranged marriage, common in India. I don't recall the details, but I felt the unmistakable closure of a lock and key about my heart. As I understood, or perhaps misunderstood, Joe was betrothed.

The warmth of his gaze remained as I descended from my barstool and wobbled Weeble-style for the door. In the bright sunlight of the parking lot, Joe's eyes remained dilated as we said our goodbyes beside my car, and I thought of the 5,500 sunsets he'd taken in from the other side of the world.

The next day, I burrowed in the silent tiers of the reference section of Parks Library on the Iowa State University campus. Often, the heavy tomes were beyond my strength to lift, so I'd slide them to the edge of the shelf, hands gripping the heavy bound volumes, and ride free fall with them to the concrete floor, where I read the articles that I'd worked so hard to collect. After hours of searching and reading, I'd turn back to survey the mess I'd created, volumes of journals scattered about like the poltergeist scene in *Ghostbusters*.

On the third floor of the library, I pressed the button to call the elevator. As the doors slid open, Joe appeared in the darkened space, the expression on his face of someone who had just spun three sevens playing the slots.

"What?" I said with genuine wide-eyed surprise. "How can this possibly happen? There's something like 25,000 people on campus."

"I have no idea," said Joe, shaking his head.

Had I looked a little closer, I wondered if I would have noticed Joe's nursing a stitch in his side. Perhaps it was just chance we met up I told myself . . . or was it? Many years would pass before I looked back on the experience and wondered if Joe had spied on my movements that day in the library and then made a mad dash down the staircase and onto the elevator to surprise me on the third floor. Maybe I'll never know.

The entryway of the Great China restaurant was otherworldly—a passageway between the brilliant sun outside and the turquoise-hued walls and beaded doorframe murals suggestive of a koi pond brimming with orange-scaled fish. In the solitude, a giant grinning Buddha stood guard, his girth ballooning into the cavernous space, standing watch beside a decommissioned cigarette machine and several Lions Club candy dispensers. A party of four exited. The metallic door clinked. A thin line of light hinted of an Egyptian mummy entombed in a sarcophagus. It was the closest suburban Des Moines came to the exotic.

I enjoyed a fantastic lunch with Joe. We were seated at an impossibly small table for two, our eating space, the diameter of a chessboard. I hadn't laughed so hard in a very long time. I was acutely aware that I was living vicariously through the life of another, but my "pretend date" felt so authentic that I found myself having a great time.

As we exited the Chinese restaurant, Joe prompted me to pause in the darkened space. As he made a move to hug me, a flood of senses invaded this millisecond of indecision—I heard the furious scribblings of an intern's notes, felt the grit of Velcro restraints biting into my four-year-old limbs, and smelled the noxious fumes of a mask of anesthesia placed over my nose and mouth. I raised my arms instinctively to block his approach.

"You don't hug; you push," Joe explained.

I've been hugged innumerable times, of course—a quick entanglement of limbs and a separation that couldn't come soon enough. To be held was a foreign concept. Joe and I channeled the kinetics of Mr. Miyagi and Daniel of *Karate Kid* fame, my arms blocking Joe's moves with each attempt.

Our quiet was punctuated by a burst of sunlight. A large party entered the entryway. Moments later, I was discovered barricaded behind three candy dispensers, huddled close to the Buddha statue. I retreated farther, but Joe drew me out.

"Arms," Joe reminded me as he inched closer to me. In my mind, I was four years old again. *Be a bird. Be a bird.* The words I can't escape. This time, I was not asked to hold out my arms like wings; I was told to rest them against my sides.

The nerves in my arms buzzed like downed power lines. Swiftly, Joe wrapped his arms around me. Instinctively, my head bucked, but his palm secured it in place. My arms fought for release like tangled limbs in a straitjacket. He held me tightly until I weakened, my arms sliding to his back, inches apart from each other.

I had attempted to master fate with my education. I could draw each of the twenty amino acids by memory, predict biochemical reactions by heart, and identify my family's gene mutation from three billion letters of code. But if I couldn't master closing a two-inch gap between myself and another, I would never, ever be OK, and these calculations will have been for naught. Moments later, my palms met.

Like my mother, Joe had the uncanny ability to register my mood in the first beats of a conversation or in a momentary read of my face. As we settled into plastic chairs on the deck of the coffeehouse, Joe leaned in closely, the brim of his omnipresent baseball cap casting a shadow over his face, drawing me closer to his world. The timbre of his voice could drop my blood pressure by twenty points. If I were to receive bad news, it was Joe I'd want to deliver it. His soulful words could melt chaos to cream.

"So . . . ," he began, his open body posture an invitation for me to speak.

I shrugged my shoulders and bit my lower lip.

"Look, something's bugging you," Joe said. "You've peeled the label off your soda into tiny strips and now you're tapping your foot against the table."

"I had an appointment with my neurologist, and it didn't go well," I began.

Throughout my childhood, my family made annual visits to the Mayo Clinic shortly before the beginning of each new school year. I knew the ritual by heart. In the corner changing room, I'd don a thin gown and stare down the poster of the normal muscular system on the cubicle wall—the cartoon depiction, skin pulled back to reveal bundled red fibers like meat in a butcher's case, as if I could magically heal myself through osmosis. Outside the cubicle, my gown was tucked into my underwear to provide a clearer view of my gait. Then I'd stand like a mannequin on a pedestal as my neurologist moved about me dictating my abnormalities into a handheld recorder.

After fourteen years of visits to the Mayo Clinic, my family had not received a precise diagnosis as to our type of muscular dystrophy. A patient coordinator from our local Muscular Dystrophy Association (MDA) office called my mother and suggested we were long overdue to check out the services provided by the MDA. Though I wasn't told this at the time, our private health insurance had suggested we check out MDA rather than the Mayo Clinic. Previously, they had been supportive of our visits to Mayo, even reimbursing travel costs.

The neuromuscular clinic visits were promoted as being on the frontier of scientific advancement and poised to churn out cures in short order. Excited by the idea of getting somewhere with diagnosis and treatment options, I shared my mother's enthusiasm. Finally, we were assured that we would meet the "people who were working on it."

By the age of twenty-six, I had forged feelings of hope and chagrin as I received my yearly postcard notice to attend the traveling neuromuscular clinic. It notified me of my date and time of appearance, much like a jury summons, without regard to the fact that I might have work or social obligations conflicting with my preassigned appointment. Though I'd dress to impress, my questions were often interrupted with "We don't have time for this," an odd rebuke that never ceased to irk me. The hours-long appointments were filled with therapists I hadn't requested—occupational, physical, and respiratory—and the punitive air that wafted in and out of the room with each consult left me drained emotionally and physically.

"'So,' the doctor said, 'why do you still live with your parents?' and not in a gathering information kind of way. She looked at me with this sneer, like I'd done something wrong," I explained to Joe.

As a graduate student in the Department of English, I lived off the pauper pay of a teaching assistant. After my car payment, I had little disposable income. Living with my parents was a logical way to afford my education, and I enjoyed their company. Her suggestions—move into a government-subsidized apartment in downtown Des Moines, rumored to be infested with bedbugs, or rent space in an elderly woman's basement—left me with the expression of someone who'd just drank soured milk.

"Live in an old woman's basement?" Joe asked with incredulity as he spun his flip phone in his hands. "Doesn't she get that you have trouble with stairs to begin with?"

At the time, I was not familiar with the euphemism *placement*. I felt belittled, and I didn't have the life experience to understand that the physician's jab regarding my living arrangement and her insistence I move elsewhere was not appropriate dialogue. I assumed the advice provided by my doctor would jibe with my educational and life pursuits. I swallowed hard, the line of questioning going down as easily as a chicken bone down my throat. I was too naive to understand that the grants to support the neuromuscular clinic, including those from the government, may have depended on adherence to a prescribed social agenda.

Joe asked me, "Why can't you meet someone, get married, and start a family?"

I carefully considered Joe's point. The agenda of a neuromuscular clinic brought me down. Talking things out with Joe provided a much-needed "second opinion" on my life goals.

"So wait, wait, wait," Joe chuckled. "According to this doctor, the entire Indian race is socially maladjusted because adults often live with their parents."

"Apparently so," I suggested with a newfound smile.

"I'm twenty-six, and I live with my mother. And you're twenty-six, and you live with your family."

We both paused for a few seconds.

"So why do you go to these appointments?"

It was the epitome of questions because it forced me to examine the reality of my medical journey, and I didn't like the patient I'd become. As a young patient, venom pulsed in my veins—I'd spewed pink medicine into the face of a nurse who grasped my face too tightly, thrown myself to the floor of Mayo's phlebotomy department during a tantrum, and scowled when asked by the third resident if I had trouble with zippers and buttons. (I'm nearly fifty, and I don't have trouble with zippers or buttons.) Somewhere along the way, I'd lost my spirit. I'd perched silent as a husk on the exam table as my parents and doctors volleyed questions over my head. Joe's questions were the first to pair "why" and "you." I felt the quickening of my autonomy resurrect within me.

However, forging this new path was not without heartache. It required me to realize that the fever pitch of the Jerry Lewis Labor Day Telethon was a regrettable mirage—that a team of "elves" were not crafting a personalized cure. Behind the plea to "just make that call" and the insistence that the "cure was just around the corner" was the grim reality of cold, plastic protractors pressed into skin and the pushing and pulling of strength testing—a system more reminiscent of an intake interview at a nursing home than a state-of-the-art bioengineering start-up. Regrettably, the neuromuscular clinic, as I experienced it, was a place where there was no healing, only documentation.

As I neared completion of a tutoring lesson with Ben, one of my third graders at the tutoring center, I couldn't help but sneak a peek at the wall clock. I had fifteen minutes before my Match.com profile would go live. Maybe I could get home to take it down in time.

As if he could read my mind, Ben asked about my family, and I explained I didn't have any kids.

He pressed on with his worksheet but looked up again at me with a puzzled expression.

"Well, you look like a mom," Ben said with certainty. "Like the nice kind, you know."

His comment got me to thinking about my experience at Target. As I circled Target on the large red scooter designed to fit someone three times my size, I'd hear the refrain, "Hey, she can't be on that thing!"

Typically, the parent, usually a mother, would burn red with embarrassment as she tried desperately to silence her curious child. After this had happened a number of times, I lingered in the adjacent aisle, letting the parent–child conversation play out so I could determine what the child meant.

"Well, Katie (or Rylie or Max . . .), sometimes people need something to help them get around, like a cane or crutches or a wheelchair."

"No, you're not listening!" the child would often interrupt. "She can't be on that thing cuz she's a mom."

In this brief exchange, I was granted the answer to a question I'd posed to myself for years. Was it right for me to dream of being a parent when I was slower and weaker than nearly everyone else my age? I could pose infinite questions to myself on the subject, but it was the straightforward style of preschoolers that cut right to the chase. If little children imagined me as a mother, then why couldn't I be one?

My fear of viewing my Match.com profile decreased just a bit as I drove home across town in the afternoon rush of traffic. Once I had returned to my apartment, I logged on to my computer and immediately covered my face with my hands. Peering through my fingers, I saw that my profile had received 6,700 views and there were sixty-seven introductions from interested men. For a moment, I wasn't sure if I was happy or horrified..

A few weeks earlier, my dad and I had a talk. I had insisted that the likelihood of a guy's being interested in me was about one in a million.

"Some people have baggage; I have an airport terminal," I would lament to him.

"It's far more likely than one in a million," he insisted, "but it is a numbers game. You have to get out there, put yourself in a target-rich environment."

Finally, after years of soul-searching, there was a black-and-white answer: one in one hundred men were open to dating a woman with a physical disability. Instead of being concerned about the ninety-nine who might say no, sixty-seven introductions could easily fill a social calendar for the next year.

One by one, I took to reading subject lines. To my surprise, the introductions were courteous and thoughtful.

"*Snakefood!*" My pointer finger rocketed toward the reject button, but at the last moment, I paused and opened the Match.com profile. I was thrilled to see that instead of a dim basement with aquarium tanks and reptiles, I was treated to a handsome guy in a white T-shirt and jeans, posed kneeling on the ground with his dog.

This was promising!

I responded with a "hello" to his profile, and we made a date to meet at Gray's Lake, a popular meeting place south of downtown Des Moines with walking trails and a beach.

On July 23, 2003, I parked my car in the lot and casually checked out the many bikers, joggers, and pet walkers passing by. A handsome man crested the small hill, and I hoped this was Jeremy. Fortunately, it was. After a brief hello, he helped me lift out and assemble the scooter in the trunk of my car, and then we started our walk around the lake.

"You don't seem shy at all," I said to Jeremy.

"I'm trying really, really, hard," he insisted, and I smiled at his warmth and honesty.

As we circled near to where we'd started, Jeremy ran over to his jeep and brought out a dozen red roses for me. We selected a picnic table and enjoyed conversation as the sun sank low in the sky and the lake water rippled with color.

"So what's the blue oval on your finger?"

Jeremy flushed with embarrassment.

"It is . . . uh . . . it was a snake head, but I got the tattoo when my finger was swollen and then it went down."

"I'm sure there's a story," I suggested, now eager to get to the bottom of the "Snakefood" profile name.

"This was back when I was eighteen. I was a camp counselor, and I got bitten by a rattlesnake, a pygmy rattlesnake."

"How did that happen?" I can't help but ask.

"I picked it up and was carrying it around for about a half hour."

It was the perfect story to break the ice on a first date. Jeremy helped me load the scooter in my trunk as if he'd already done it a million times. We made plans to meet for a canoe trip down the Boone River on the weekend.

I watched the swift waters of the Boone River with much trepidation as I sat cross-legged on the shore surrounded by a dozen intrepid twenty-somethings. A canoeing adventure with an 8:00 a.m. start time was not my typical Saturday-morning activity. Despite my fear, the morning was infused with a brisk awakening of my senses, though I eyed the steep embankment to the water wondering how I would possibly navigate getting into a canoe.

On this occasion, I didn't have to worry. A man of few words, Jeremy had an air of confidence that put me at ease. Without hesitation, he gestured for me to clasp my arms around his neck, and seconds later, we were in motion down the embankment toward our waiting canoe. Instinctively, I clung to Jeremy as the water pooled thigh high about us and we got into the canoe. With a swift push, we were propelled into the current, our small aluminum craft gently rocking back and forth.

I hadn't traveled by canoe in nearly twenty years. The last time I'd enjoyed the experience, I was a nine-year-old at Girl Scout camp. The experience allowed me to simultaneously travel back in time and dream of a future. Without Jeremy, I couldn't have possibly navigated this adventure on my own.

Ninety-plus degree temperatures were predicted for the day. I'd broken with habit and donned a bikini top for the trip, but I'd opted for jeans to hide my skinny legs. I wore a cover-up, but as the noonday sun beamed in the crystal blue sky, I regretfully removed my shirt. Then it dawned on me that I hadn't shown my arms to anyone but family since I was a young girl.

As we moved through the waters, I realized that Jeremy, seated behind me, could take in my long scoliosis scar, another fragment of me that almost no one had ever seen. A half hour into our journey, Jeremy moved closer behind me and mentioned my reddened skin. Would it be OK if he applied sunscreen? As he worked the lotion into my skin, I couldn't help but wonder what he might be thinking. I hadn't offered any explanation for the scar, but his hands worked methodically, massaging my skin as if he'd done it dozens of times.

As Jeremy continued, he told me about his diagnosis of spinal curvature. Unlike my scoliosis, which altered far more of my spine, Jeremy had been diagnosed with mild kyphosis, a deformity of the top part of the spine.

"I had to wear a brace at night, and I absolutely hated it," Jeremy said.

It was the last thing I'd ever expected on a date. I was so worried about navigating what had happened to me that I hadn't thought about my date's sharing a diagnosis.

When we returned to my car, I had hoped we might make future plans, but Jeremy made a quick exit back inside his office, our meeting place for the canoe trip.

"See ya!" he said as he walked away.

I sat in my car a few moments before realizing I couldn't find my sunglasses. Had I left them in Jeremy's jeep? This was my chance for one more connection. I called Jeremy, and momentarily, he returned to the parking lot, scouring his jeep for the glasses.

"Nope, sorry," he said. "See ya!"

On the drive home, my mother called to learn more about the date.

"Oh, he's a great guy, and I had a lot of fun. But there won't be a second date," I shared.

"He said, 'See ya,' so I definitely know he's not interested."

She offered to stop by my apartment and check out my Match.com inbox, and an hour later, we were sipping strawberry milkshakes and reading through profiles. Momentarily, I checked my email.

"Well, I guess I was wrong about this," I said as I pointed out the new email from Snakefood. "He's inviting me to a movie for our next date."

Seabiscuit led to a first kiss and a lost car in the theater parking lot, which led to the hot-air balloon races in Indianola, which led to dinner at Fazoli's and an evening drive around Saylorville Lake.

By early August, we found ourselves at the heart-pounding, larger-than-life spectacle of the Iowa State Fair midway. I lived ten years in a single night, clutching carnival booty tightly to my chest as Jeremy walked up and down the rows of carnival games, taking entirely too long to decide which to go for.

"What's taking you so long?" I asked.

"I'm trying to find one you can play," he said.

My eyes filled with tears.

"This is the happiest I've even been!"

8

DELIBERATIONS

"How do you imagine your son feels about your decision to have him despite the genetic risk?"

The 2,200 pairs of eyes focusing on me increase the intimidation quotient of the question a hundredfold. My pupils enlarge in a futile attempt to take in additional light as I scan the inky black darkness of the Greater Columbus Convention Center in search of a disembodied voice. With the glare of the bright lights, I can only make out the first two rows of attendees, most notably, three ladies whose facial expressions sour as they turn their heads back in the direction of the microphone.

So it isn't just me.

Emotion fills me in a kind of my-life-flashed-before-me way, but in this case, my love for my son is center stage. It's a question I can't imagine fielding from my best friend or my pastor, let alone a stranger at the 2017 National Society of Genetic Counselors annual conference. However, I promised a candid conversation when I accepted the opportunity to provide a keynote speech before half the genetic counselors in the nation. Perhaps the question is fair game.

I didn't have the same odds for genetic risk as the average mother when I conceived my son in 2005. For nonrelated persons, the odds of having a child with a birth defect are approximately 3 percent of pregnancies. For first cousins, the risk doubles to around 6 percent.[1] As a person with a dominantly inherited genetic disorder, any baby I conceived naturally had a fifty-fifty chance of inheriting my rare form of muscular dystrophy.

Chapter 8

I'd long since recognized myself as the elephant in the room at baby showers and having the feeling of wanting to be elsewhere with the "boy or girl it doesn't matter so long as it's healthy" talk. Don't get me wrong—I'm all for healthy babies—who isn't? But there's the inevitable self-deprecation of hearing your lone identity singled out. Try any other characteristic, such as so long as the baby's not short, not freckled, not knock-kneed, not amber haired—you get the point.

When my fiancé, Jeremy, and I met at the county courthouse to apply for our marriage license, I arrived with newly coiffed curls from the salon and a sunburst T-shirt dusted with glitter. I'd wanted to mask my dis-ease with the whole process. We were required to bring a witness, so Jeremy invited his jovial coworker Kevin, who snickered as he was asked to certify that we were, in fact, not first cousins. I'd feigned a giggle, but inside, I was troubled.

It's not that I think marriage between close relatives is free of concern; it's that I consider the county clerk's office an unsuitable setting for that discussion. I agree that it's an important issue and that it should be explored in a physician's office or in counseling with a religious leader or genetic counselor, not probed alongside requests for boat trailer renewals and applications for vanity plates.

Behind this seemingly benign question was a purpose—to weed out marriages having increased genetic risk. As a lifelong Midwesterner, firmly anchored to the buckle of the Bible Belt, I knew the prevailing sentiment of Iowans. I would dutifully reply that we were not first cousins when questioned, all the while possessing the knowledge that any naturally conceived children of ours faced a far greater statistical chance of genetic illness than anyone proposing to marry a first cousin. With no one asking "Will there be genetic risk to this union?" we slipped past undetected when others with far less genetic risk might be turned away. What would become of a couple who announced they were close relatives? Did a large stamp labeled "invalid" come down on their license? Was the sheriff called out to investigate? Suppose the closely related couple planned to utilize a donor egg or sperm to achieve pregnancy and reduce genetic risk or preferred to share a life as a couple and not have biological children. Would the license be denied to them as well?

To this day, I can't testify that I made the right and moral decision to have my son. A person could assign selfishness to my life story, and with barely a blink of my eye or a rise in blood pressure, I would simply say, "You may be right." This is not to say that having my son wasn't the most awe-inspiring and rewarding experience of my lifetime—that not a day goes by in which I don't marvel at his six-foot-tall frame and wonder how this glorious gift sprang from my broken body. On the other hand, my heart never ceases to grieve for parents who fall on the other side of this fifty-fifty coin flip, those who must daily question the rationale of their decision when seeing their child's suffering.

My mouth grows parched and my eyes water as I sift through twenty-five years of indecision. This ethical dilemma has been the ultimate question of my lifetime, and I have no answer to satisfy this counselor's curiosity. Instead, my mind races back to a kindergarten memory, the day the maddening contemplations began and the sound of a china doll face splintering on the ground—a sound that I can still recall forty years later.

1980

The tiny crawl space between the sofa and the living room wall was the perfect place to hide my kindergarten body as I snuck out of bed to watch *Saturday Night at the Movies* while my mother pieced together a latch hook rug, certain that I was sleeping soundly in my bed. The plot of the TV movie was way over my head. A teenage girl is hugged and kissed by several football players in a locker room. The coach finds them and is angry, and the girl runs away. In a later scene, the girl is crying as she is wheeled out of surgery on a gurney. Her mother greets her warmly and presents her with a china doll with painted lips and rosy cheeks. The girl is filled with rage at the sight of the doll. She rips it from her mother's hands and hurls it to the ground. All I can hear is the sound of shattering china echoing in my ears.

"How could you do this to me?" the girl questions between sobs.

I listened intently to their conversation. One point was crystal clear. Whatever happened during the surgery has changed her life forever— she can no longer have a baby.

I rose from my hiding spot with an audible gasp.

My mother was startled, but she didn't scold me for sneaking out of bed.

"Would you do this to me?" I asked.

I could plainly read the horror on her face. Not only had I conceptualized the movie's plot, but I'd also connected with it.

"It's not a movie for children," she insisted. "You should be in bed."

"Would you make it so I had an operation so I couldn't have a baby?"

"Oh, no!" she gasped. "I would never do that!"

In a clumsy conversation, my mother attempted to ease my worries. She explained that the young teenager had an intellectual disability that made it difficult for her to handle a pregnancy and care for a baby.

"You're a bright girl," she reassured me. "You would know how to care for a baby."

My brow furrowed in deep thought.

"The mother, maybe she felt she was doing the best thing in this situation," offered my mother without conviction.

"Then why is she crying?"

The words hung in the air as I was ushered back into bed.

"Were you mad at your dad for having you?" a friend once asked me as we strolled into elementary school together. She'd noticed that I walked with the same unusual gait as my dad.

I stared at her in bewilderment, as it was such an odd question. It was like asking a baby bird "Why do you peck out of your shell?"

"I wanted to be born," I replied without hesitation. At the age of eleven, this direct question evoked a visceral reaction, and there was only one response.

By thirteen, I'd developed a more nuanced point of view. Maybe it was the tightening of the grip of the disease on my body or a more developed social consciousness, but often, I found myself in deep thought. My attachment to the physical world lessened, and I doggedly pursued an understanding of my spiritual self—one foot in this world and the other heaven bound. Assuming there was a finite amount of suffering in the world, why did some bear the brunt of grief so deeply as others appeared perpetually blessed? Could suffering be diluted? Bartered? Freely exchanged?

A favorite pastime with my mother was the "Would You Rather" game.

"Would you rather be covered in honey and sit on an anthill, or would you rather climb in a bathtub filled with snakes?"

Each time, her answer was the same: "Well, neither."

To which I'd reply: "You have to pick one."

"I suppose the snakes," she'd say as she squirmed in place.

For my mother, it was simply a silly guessing game, but for me, it was a stress-releasing exercise, a way to transform suffering into quantifiable puzzles.

Though talking it over helped to ease my mind, at times my frustration was overwhelming. On one occasion in the seventh grade, I became upset on a car ride home from school, and my mother pulled the car over to the side of the road. Though the reason for my tears is lost to the passage of time, I remember asking why I couldn't have died as a young child—why I couldn't take the place of another parent's healthy child lost in a car accident. Why was I here if I'd entered the world so flawed?

"I wish I weren't here. I wish I lived in heaven," I suggested between sobs.

My mother's steepled fingers tapped her nose and mouth gently as she exhaled slowly.

"This isn't normal," she said quietly. "You shouldn't feel this way."

The words in my head throbbed—maybe this was *my* normal.

Kindness and comforting words followed, though I can't recall the details. We groomed ourselves to hide the hurt—dabbing red-rimmed eyes with Kleenex and smoothing wayward strands of hair. It was an unspoken rule that we never showed this release of emotion in front of my dad. We didn't want him to feel sad.

On a visit to the mall at age thirteen, my muscles began to quiver with exhaustion. I took a seat on a bench next to a woman with long curls of raven-colored hair. We exchanged brief hellos and studied the passersby.

Her silence stirred my memory, and I realized that it was not the first time I had taken a seat beside her. We had met like this on several other occasions. I found myself wondering what plucked her from her shopping and led her to this bench as the cramping in my legs had led me.

Moments later, my mother and aunt arrived, and we crossed the parking lot to our car.

"The woman on the bench beside you, did she say anything to you?" asked my mom as she shut her door.

"Not really. We just sort of said hi," I offered, perplexed by the question. "Is she someone we know?"

"She's just someone most people recognize," my aunt said. "You're too young to know, but most adults know who she is."

"Well, so who is she?" I asked, annoyed with the evasive conversation.

"She's Noreen Gosch," my mother said. "She's Johnny Gosch's mother."

I needed no further reminding. Though I'd been only seven years old at the time of the abduction of the twelve-year-old paperboy in West Des Moines in 1982, the idyllic years of elementary school would never be the same. Assemblies were held as safety officers flattened our inked palms on identification cards and we learned the rallying cry "Run!" if we ever found ourselves in danger. I recalled sinking low in my seat as my friends leapt up around me and wondering what I was supposed to do since I couldn't run.

Many years later, during a visit to a Coldwater Creek store, I juggled my toddler-age son on my lap as a store clerk handed me an orange juice in a small plastic cup. She placed donut holes in the napkin cupped in her hand, then set it on the table before me. I hadn't had the chance to thank her, so I put my son down, eyeing his movements as he zoomed about the tables of crisply folded shirts and stacks of sweaters. It was only when I took note of her name tag that I realized it was Noreen Gosch who was serving me. I blinked back the tears in my eyes and swallowed hard to clear the lump in my throat. I couldn't help but wonder if the sight of a young mother and son tugged at her heart. Did the question of "why me?" ever fade in intensity for her? How did she face this day—rising from her bed and coming to work to serve another? Why was I the one holding my son on this day?

Over time, my game of "Would You Rather" had transformed into an unraveling of the meaning of suffering. Who had the hardest life? How did one persevere against all odds? Did a hard-hearted God study life from the heavens, granting yes and no to prayers for intercession? These questions would boggle my mind for hours. With maturity, I'd realized the parent of an abducted child rose to the top of the hierarchy of human suffering, racing past those with disabilities, separated couples grieving divorce, and those who'd lost entire fortunes to drugs and gambling debts. I'd come to understand that the wounding of the heart was a pain so unsolvable and that no one grieved more than the parent of a missing child. I had uncovered "the hardest life."

One morning, I awoke to find my dad studying baby photos of me that he'd found on my bedroom closet shelf. He suggested I go back to sleep, but my curiosity was piqued—why the sudden fascination with my baby photos? We were weeks past my high school graduation. The Tour de France of childhood was over, and I would be leaving for college in the fall. It was time to pack away my emblems of the past, not stare at them in reverie.

Perhaps my dad was a closet *National Geographic* photographer. He had a knack for contorting his body into impossibly tight nooks and crannies for the best angle to take a photo. Our family albums contained hundreds of photos, many captured from a child's vantage point—a photo of me lying prone on my bed, a grin to the camera; a profile of me, deep in thought, gazing into the fish aquarium and rippling water mirroring on my face; and his all-time favorite—a close-up of my bare feet as a focal point on the diaper-changing table.

"Oh, the places you'll go," he'd coo, Dr. Suess style, when as a child, I'd sit on his lap turning the pages of the book.

This morning, he held photos of me at arm's length, placing each against a trifold sheet of pastel paper, then turning the image vertically and horizontally as he searched for the perfect fit. I would later realize he was seeking photos to support a pro-life advertising campaign. With my picture, he was plotting the money shot—the kind launched on a billboard over a cornfield or pictured on a pamphlet stuffed into the back pocket of denim jeans worn by a frightened teenage girl.

In his mind's eye, I was a baby-faced bumper sticker—"I'm a child, not a choice"—that materialized from a plume of smoke from an exhaust pipe. My merit for living was being spelled out in the knowledge that I began to develop fingerprints[2] in the first trimester of pregnancy and had tiny earlobes at nine weeks' gestation.[3] It was my face, he envisioned, that could compel a pregnant woman to make an about-face at a clinic entrance. Me, surrounded by words like *Precious*, *Life*, and *Jesus Saves*, thrust in the air on white poster board. The trouble was, I didn't know if I belonged there.

As a preschooler, I was naive to my father's cause. I'd accompany him to the Craftsman-style house in the heart of Des Moines, indistinguishable from the others in the neighborhood except for the "Iowans for Life" sign, flanked by two long-stemmed roses. I didn't know anything about the political machinations that orbited the home. I just knew it as the "house that loved babies" and by the Snookies Malt Shop next door. Inside, I'd circle the rooms, gazing at the posters of the babies on the wall, taking in this new world order where babies didn't cry, didn't teethe, and didn't need. Seated on the Xerox machine, I felt the hum of progress as papers churned out and were sorted into a pile, dozens of baby faces peering back at me, still warm from the press.

As a high schooler, I once volunteered to stuff envelopes for the organization. As I perched on the edge of my folding chair, a radio broadcast summarized the legal motions up for a vote throughout the country. The air was thick with anxiety, as relaxing as a picnic lunch on a guillotine platform.

The truth was, I was uncomfortable because I didn't know if I would ever carry a baby. By all accounts, it was a long shot considering my five-foot three-inch, eighty-seven-pound frame. Wasn't there that statistic in every biology text citing 120 pounds as the magic weight for fertility? I felt like I didn't belong, like the girl with no rhythm who is assigned to pass out Gatorade to a flock of pelican-legged dancers.

The relatively new field of preimplantation genetic diagnosis[4] provided options new to my generation. During the process of in vitro fertilization, a single cell could be siphoned from an eight-cell embryo (blastocyst) and tested for a known genetic mutation. In my case, the location of my G to C alteration had been identified in 1999.[5] Scientists could search for this precise change like a beacon among three billion letters of genetic code. If the mutation was not present, the blastocyst could be implanted in me and would, hopefully, result in a pregnancy. However, at the time, the technology was considered experimental and was not covered by insurance in most cases. When I investigated my reproductive options, the cost carried a base price tag of $18,000 for a single in vitro cycle, including preimplantation genetic diagnosis. In addition, the quoted rates of success were half those quoted to couples without genetic screening because half of created embryos would carry my genetic mutation.

In the spring of 2000, at the age of twenty-five, I made an appointment at a local fertility clinic. I wanted to learn more about this revolutionary option, and I requested a consult with a fertility specialist and a blood draw to determine if I could possibly conceive a child. There was no immediate need for the visit; I had never even had a boyfriend.

My name was called, and I followed the nurse into a cozy office. The doctor rose to shake my hand, and I sat in a plump leather loveseat, thankful for the posh surroundings rather than the thin gown and paper-lined exam table I'd been expecting. Pink-hued models of male and female pelvises hovered on the shelves beside me. Suddenly, I felt out of place. I didn't know anything about love. I had never even visited a gynecologist. This was a place where the miracle of reproduction had lost its wonder, a place where those who had tried nearly everything now turned to this doctor for a miracle of science instead. I didn't belong here; my eyes searched for the door.

"You're a very normal twenty-five-year-old woman," the doctor began as he opened my medical file.

The word *normal* and my medical file had never occupied the same conversation space. I shifted uncomfortably in my seat and gazed out the window at the bustling parking lot filling with patients.

"Your hormone levels, including follicle-stimulating hormone, are all in normal range for a woman your age," the doctor explained as his clasped fingers rested idly on my file.

The dark humor of this medical revelation overtook me. I couldn't begin to count the number of ways I was different from the typical person, but here, I received a clean bill of health.

It was my brother Bob who made me reconsider my thoughts. Seated across from me in a dimly lit pub, he listened intently as I described the miracles of modern-day genetic technology with numerous hand motions and dramatic flair. I was effervescent with joy that the worry of genetic fate could be banished from our family line, perhaps forever. Leaning into the glow of the bulb above us like a captive in an interrogation room, he responded: "So what if Mom and Dad had done this?"

For a few moments, I pictured my dad and siblings, and friends with Emery-Dreifuss muscular dystrophy, and in Bob's eyes, I saw a reflection of future children. If he declined in vitro screening and I pursued it, how would I reconcile feelings of fairness and family loyalty if some of the children in our family inherited EDMD and some did not?

In the weeks that followed, I pondered the results of my fertility clinic visit. One night, I was awakened at 4:00 a.m. by a phone call. As I placed the receiver to my ear, I heard the excited voice of Dr. Luciano Merlini, a physician from Pavia, Italy. I recognized his name as one of the physicians in the original genetic study that found the gene for my type of muscular dystrophy. He apologized profusely for the confusion regarding the time difference.

"I thought you would want to know . . . there's a woman like you, one of my patients, she has a child, a son!"

Immediately, I was jarred awake. This was such good news!

"But I'm so small," I began. "I can't imagine this could be possible."

"Yes, I've seen photos of you, and I can assure you that this woman is very small as well," Dr. Merlini continued.

In the conversation that followed, I learned of this woman, so much like me, living halfway around the world and even sharing my exact mutation. He told me of her father, a man in his sixties who walked with a cane, and her teenage son, also affected by EDMD but to a milder degree than his mother.

After the call, I settled back onto my pillow. I had forgotten the most important question—I did not know their names. Gabriella, Mario, and Antonio, I whispered to myself. My heart swelled with joy as I pictured this family taking in a glorious sunset on their veranda overlooking a vineyard of tangled grapes. I heard the familiar tap of my father's cane as this family and my own mingled in my dreams.

"There may be one bride in ten thousand thinking about genetics as she walks down the aisle," barked Dr. Laura, a conservative talk show host known for her biting commentary.

The evening's caller was a young woman who was getting married in a few weeks and was despondent because she had learned her adoption records were sealed. Because of that, she did not know her health history and felt unprepared to start a family. However, Dr. Laura was not swayed in her opinion that adoption records should remain sealed.

"People always want to make a big deal of genetics, but the overwhelming odds are that everything will work out fine," Dr. Laura insisted with finality.

Talk show music played over the caller's tears, and I snapped off the radio. My windshield wipers punctuated the silence like a metronome. As I stopped at a four-way intersection, torrents of rain obscured the identities of the drivers of the other cars. Part of me preferred the anonymity to ponder the radio show, but part of me felt more alone than I ever had before. I was twenty-five years old, the same age my mother had been when she had me, but I was nowhere near ready to make peace with my flawed genetic inheritance.

I've never had the ability to mask my emotions, and with one look in my direction as I entered the front door, my mother knew something was wrong. I broke down and cried on her shoulder as I regurgitated the gist of the call and the advice given.

"Dr. Laura said, 'One in ten thousand brides think about genetics as they walk down the aisle,'" I sputtered.

"I did."

My mother's words were barely audible. I pushed my body aways from hers, studying her with unblinking eyes. I didn't know if my eyes conveyed anger, hurt, love, or all of them.

"I thought about it," she said.

This was the first time my mom had ever spoken to me about this. Throughout my life, the emphasis was always age four—the certainty that this was the age when I began to show weakness, which became the guiding storyline of all conversations with a physician. With this revelation, I found my curiosity piqued regarding my life before age four, as if the earliest chapters of my life story had been ripped from a diary and I had only the shreds of the bound pages to assimilate my past. *Was this lie of omission intended to help me? Was it for my mother's benefit? What could be gained by kicking the can of reality further down the road?*

Picturing my mother as a young bride was both enchanting and foreboding. In the idle time between Sunday school and the church service, my five-year-old feet would retrace my mother's steps, traversing the secret staircase from the bridal dressing room in the basement to the back of the sanctuary of First United Lutheran Church in downtown Des Moines. In my memory, the steps were draped in plush red velvet carpeting, and the golden spiral banisters were flanked by flame-lit lanterns. On my return visit forty years later, I was disappointed to find faded threadbare carpeting of indeterminate hue and dull wooden dowels as railings, softened with the passage of time.

My adult eyes couldn't overlook the all-encompassing aloneness of the path, as the twisting stairs allowed for the passage of only one—the concentric steps bearing the unmistakable analogy to the double helix of DNA. My mother could not have ascended the stairs quickly for fear of stumbling, and I wondered at her thoughts as she made these blind footsteps, her gown masking the floor. In this short passage, she departed the comfort of her closest friends and emerged on the next floor, guided by her father's hand, into the candlelit sanctuary.

How did my mother reconcile the odds of the births to come— four affected children and one healthy child? How did she reconcile this with her faith in God? Did she feel betrayed?

In a bustling fast-food parking lot, I removed my birth control pills and lay them across my dashboard. Closing my eyes and pressing my forehead to the steering wheel, I sought the counsel of a future child—one without gender, face, or name.

I could change my mind, I reminded myself. I could put the pills back into the monthly packet and save this maddening contemplation for another day. Then I was interrupted.

"Do you want to try our chicken poppers, chili fries, or Strawberry Explosion milkshake?" prodded the overly cheerful waitress on roller skates. She leaned so closely into my car window that I could smell her breath mint. I felt painfully exposed, and in a single swoop, I collected the pills in the palm of my hand.

Her interruption overwhelmed me. I thought of the mundane choices I made every day. My mind was weary with the one choice that meant everything, the choice I had pondered for twenty-five years: Do I conceive a child when there is substantial genetic risk?

I dismissed the waitress and set off toward the entrance for the highway. Out of three million letters of genetic code came a single change from a *G* to a *C*. I held the pills out the car window. As I gained speed, I loosened my grip, and the tiny pills pelted the side of my car.

As my arms plunged in and out of the warm soapy bubbles, moonlight streaming into the kitchen window, Jeremy leaned into me, wrapping his arms around me and squeezing me tightly.

"If you were carrying my baby, I would be so happy," he purred.

The wonder of his words was not what was said but what was unsaid—not after scientists make a breakthrough discovery, not when we have harvested eggs from an egg donor, not when we've saved enough for a round of IVF with preimplantation genetic diagnosis. I eased back into his body, rocking softly. This was the magic of melding with my love—the siphoning of fear, the foretelling of a shared future, the serenity that I was no longer alone with my thoughts.

Jeremy did not fear a naturally conceived child as much as I did. From our first date, I had shared the name of my diagnosis—Emery-Dreifuss muscular dystrophy (EDMD)—and he'd read up on it in internet searches. He knew my disease weakened the muscles involved in gross motor skills, such as walking, rising from a chair, or lifting objects over my head; however, there was wide variation as to how severely the condition would affect a future child. My case was severe on the spectrum, and I had lost the ability to walk at the age of thirty-three.

Within minutes of meeting, Jeremy had asked the most difficult of questions—*Could I die young from my genetic disease?* I explained the effects on the heart were the most severe aspect of EDMD, that over time, my heart would enlarge to the point that it would become an ineffective pump, much like a stretched-out balloon. Medications could slow the process, and the implantation of a defibrillator could be lifesaving in the event of dangerous arrythmias common in people with EDMD.

We talked about how EDMD is a slowly progressive condition—often five to ten years could pass before I noticed changes in my abilities. Fortunately, fine motor skills were well preserved. Such things as grooming and dressing and all aspects of self-care were second nature to me. Even driving was a well-developed skill.

My dad was a textbook case of EDMD. He walked with a cane in his fifties and gradually lost the ability to walk by age fifty-seven. He had been graced with the capacity to build a loving marriage, to pursue a twenty-five-year career as a federal prosecutor, and to raise five children, all the while serving as the breadwinner of the family. A regimented exercise routine built into his morning routine—down two flights of stairs to the basement shower, up two flights of stairs to his room to dress, and then down a flight once again to leave for work—likely stalled the advancement of his weakness. I could count the number of days he'd been home sick on one hand.

No, Jeremy didn't want a child of his to be born with EDMD, he'd concluded after digesting the medical information. But if this were the case, he would be filled with love for the child regardless, and he could imagine a rewarding life for the child.

As for me, I clung to the hope that maybe if my baby inherited my genetic condition, they would be far less affected than me. This sometimes occurred as it had in the case of my friend Mary and her daughter, Kelsey. However, this was an unpredictable phenomenon. Could it be more severe for my child than it was for me? I shuddered at the thought and banished the idea as soon as it surfaced.

To open my heart to a naturally conceived child required me to ask the ultimate question of my life: if I were given the knowledge that I would face a challenging disease, would I have chosen this path for me? On the surface, the answer is obviously no. *Why would I possibly want to face such an uphill battle? The humiliation of being different? Years of loneliness before I met my husband?*

On the other hand, why would I have wanted to be anyone other than my father's daughter? There were countless fathers in the world, some kind, some cruel, some impatient, some prone to abandonment. I had been so remarkably blessed, why would I imagine an alternate reality?

My greatest fear was mediocrity. Suppose I was granted the chance to see what our lives would have been like if we hadn't been challenged by genetic illness. What if I ventured into an altered reality and found that my father had never escaped the grip of poverty or hadn't been the first in his family to attend college? What if he never rose to the ranks of a federal prosecutor? What if I hadn't inherited his intellectual prowess? I was granted the experience to live as Scout, in the shadow of Atticus Finch, and though my connection to my dad carried a stiff genetic sentence, ultimately, I wouldn't have had it any other way.

As a child, I was enchanted by the film *The Secret of NIMH*. In this 1982 animated film, mice were engineered to be superintelligent after receiving an injection from scientists at the National Institute of Mental Health. Finally, there was an explanation for my father. Perhaps he'd also received an injection of bubbling liquid that weakened his muscles but also made him extremely intelligent. As I grew older, of course, I realized my logic was a non sequitur, but still, I marveled at the concept. Inheriting this disease had nothing to do with increasing intelligence or motivation, or did it?

When I was first selected as a juror in April 2005, I imagined the experience would embolden me, that I would feel a surge of power blaze through me as I offered a guilty verdict. I imagined it to be a Zen-like experience, but I found the opposite to be true—it hurts to decide the fate of another, especially one that cannot be changed or retracted. I came to understand that with this power, comes awesome responsibility and a deep lien against the soul.

To decide the fate of my child, I'd paced like a cornered tiger on a high dive for twenty-five years. Countless times, I turned back toward the ladder and counted the steps to retreat. Other times, I curled my toes at the edge of the board, stalling my next step. My vantage point was so high that I couldn't gauge the depth of the water in the pool. Finally, in a moment of sweet release, I stepped to the edge of the board ready to make the ultimate decision of my life.

Faith, my seamstress, tilted her head to the side and gazed quizzically at my wedding gown. My wedding was just two weeks away, and we were attempting the final fitting following alterations. Something was "off." I could tell by the expression on her face. She shifted the bodice ever so slightly and bit down on her lower lip.

"I'm sure I measured correctly," she insisted, after removing straight pins from her mouth with her free hand. She stepped back to take in the sight. I felt like a spring-loaded jack-in-the-box as it's freed of its claustrophobic enclosure; I was bursting onto the scene, displacing satin and sequins, and straining the seams of the gown. My once barely B-cup breasts had ballooned into C cups, all in the space of a week.

I pretended to be preoccupied with the gown, though my mind was a blur. I'd imagined a pregnancy reveal might include enigmatic moments of doubt and curiosity—is that one line or two? Is that a plus sign and a smudge? I'd never envisioned dissecting the most important development in my life by gauging another person's facial expression. Wouldn't it take months for my body to change? Wouldn't it be my abdomen straining against the taut fabric? Maybe it was all the excitement of the wedding, or maybe I was retaining fluid.

Weeks earlier, my plan had seemed foolproof—I'd thrown my birth control pills out my car window. If anything happened, it was fate, right? If I became pregnant, I could negate my fears in the warmth of my wedding receiving line, the joy of opening gifts, the well wishes of family and friends. Aboard our cruise ship on our honeymoon, I could dine on exquisite meals, partake of fruit sculptures, and relax in the warm rays of the sun.

It seemed such a preferable idea over the fearful me of an alternate reality, overcome with worry in the routine of life. I'd imagined stashing a positive pregnancy test deep in the trash and taking off alone, driving for miles with no purposeful destination in mind, then checking into an isolated motel in the countryside, my car, the only one in the parking lot, maybe a tumbleweed skirting past. Alone with a burner phone and a handful of cash, sitting cross-legged on a bed, drinking a Slurpee. Silent except for the hum of the wall air-conditioning unit. Alone with my thoughts.

I was afraid of panic. It would only take one negative look, one word of reproach, one tsk of disapproval of me as a potential mother to break me. I was fearful, and it was obvious that my worst enemy was me.

A few days later, I decided it was time to know for sure. I was so unacquainted with the pregnancy testing process that I peed with the cap still on the test wand. After downing a small bottle of apple juice, I jumped in place a few times, trying to prime my body for another test. After a barely there stream and a few moments for the wand to sit idle on the counter, I turned the test over and found a plus sign lit up like a Fourth of July firecracker. Surprisingly, I was bathed in a wave of elation. The ripple of joy in my chest felt as autonomic as the act of breathing. Twenty-five years of indecision and worry released from my body with a shudder. This was over. That was beginning.

Like any young girl, I had imagined what my wedding would be like, but I never had a clear perspective of me and my dad walking arm in arm down the aisle. It wasn't that I hadn't imagined the closeness of the day—his words of advice whispered in my ear, his sharing a favorite memory of us. Rather, it was a practical matter that had clouded my vision—we shared the same wide-legged stance, our locomotion reminiscent of a cowboy's first steps out of the saddle. In humorous premonitions, I'd imagined our legs tripping one another and our bodies hurtling toward flower arrangements and flaming candles.

Six years earlier, at the age of fifty-one, my dad had proudly walked my sister, Janet, down the aisle of her chapel at St. Olaf College. His steps were solidly planted, and he walked at a generous clip, looking every bit the English gentleman in his coat and tails.

With the passage of time, we both knew our steps were numbered. Together, we come upon the idea of a golf cart as a mode of transport down the aisle, and our florist set to work transforming the white cart with calla lilies, gerbera daisies, and ribbons of our chosen color palette—red and salmon—with hints of yellow and lavender.

My dad sat waiting in the golf cart alongside the clubhouse as our wedding guests waited meters away. Our trellis teetered with a gust of wind, and two polo-sporting caddies were recruited to support each side. Wordlessly, my dad studied my face, inhaling deeply in newfound wisdom.

"I've never seen you so peaceful," he said.

We both understood the significance of these words. To the outside world, I was the child with the ubiquitous smile, full of laughter and friendliness, but to my parents, I was a brooding and sullen child, prone to wordless tears.

Beneath my bouquet of flowers, my thumb strummed my abdomen, and I thought about this secret child, rocking gently like a baby seahorse in a gentle current, hidden from the outside world. Could this addition to our lives, barely a half inch in length, change the way the sun reflected on my face? Soften and soothe years of worry? Cause each cell in my body to leap in wonderment?

Moments later, our photographer captured the photo he will dub "the pinnacle of his career." I sparkled like Doris Day, though my legs quivered in exhaustion, and I clung to Jeremy for the strength to make it through the ceremony. There was much to fear and much to question, but ultimately, the three of us were going forward as a family.

In the summer of 2017, I visited Dr. Lori Walrath's laboratory at the University of Iowa with my son, Martin, who was then eleven years old. With a panoply of candy, Dr. Walrath explained the workings of genetics. Together, they constructed the winding double helix of DNA—intertwined licorice, protruding gumdrops, and staggered thin mints. It was the fulfillment of a promise that I had made to my future child on that fateful day in 2005 when my heart was open to conceive him. I would become versed in genetics as one becomes fluent in a second language, and I would pass the knowledge onto him so that perhaps someday, Martin could understand the choices I had made, the hopes I had for his future, and how much I truly loved him.

The journey in my mind lasted only a few seconds, and the audience waits for my response. Suddenly, the perfect answer, the only answer, forms on my lips: "You'd have to ask Martin that question. He's the only one who can answer."

9

RESPONSIBLE BIOHACKER

"So you're bedbound," said the nurse as she walked around my hospital bed and silenced my beeping IV.

My chest burned with indignation, but I could understand the nurse's assessment. Lines of plastic tubing encumbered my movements, and my mobility scooter was tucked into a corner. My tiny bird legs jutted beneath my blanket like twin deflated circus tents. I wore the thin skin and plentiful bruising of the elderly, and my nail polish was spotty at best. Like a crumpled marionette, I had fallen sideways, slumped in the contours of my hospital bed. As my final act of humility, I reached for the nurse's hand to upright myself.

I had truly arrived, I chided myself.

Bedbound. Simply reciting the word ensured a grimace. It struck me: there was no further place to fall. We'd sailed past the out-of-vogue nomenclature—lame, infirm, wheelchair-bound—and the more tolerable but awkward euphemisms—differently abled, mobility impaired—and had arrived at a place that required no further explanation—bedbound, a label that suggested I had no further aspirations for my life than what I could summon from others to bring to my bedside. I had nowhere to be, no agency in my life choices, no dreams for the future that couldn't be cordoned off by bedroom walls.

The nurse's assumption was a first for me. I had had more than forty-five years of experience living the life of a physically disabled person, but no one had previously considered me bound to a bed. It wasn't as much her word choice that irked me but rather, the implied advancement of my disease.

I had first encountered this morbid term as I planned my tenth-year birthday party in 1984. I recall slinking behind the living room sofa with my mother's book *Betty Crocker's Parties for Children*.[1] The festive book with the hot pink cover was filled with party planning advice and recipes. As my birthday approached, I dog-eared the pages detailing games for my next party. Sometimes my mother and I perused the book together, and on one occasion, I noted a section toward the back offering advice on how to talk to your children about sex. Once I had this book in my possession, I was excited to learn all I could about sex. If Betty Crocker was offering advice, I was all ears. To my disappointment, the "racy" part of the book defined sex as the number of boys and girls to invite to a party.

I'd found my answer, but I continued snooping through the parent instructions. A new topic caught my attention: "Parties for the handicapped child." *Was this me?* Yes, I concluded. This must be about me. Though my stomach turned somersaults, I read on.

"Sometimes children are confined to bed with noncommunicable diseases and are able to have a few friends in to celebrate the occasion. Children in wheel chairs may have or attend parties. Such parties should have few guests and be very short."

I snapped the book closed in horror. *The birthday child in bed?*

As a fourth grader, I had very little understanding of the nuances of medical conditions. I hadn't previously considered that a severe illness, such as a cancer diagnosis, could necessitate a scaled-down celebration. I hadn't appreciated that a child's wheelchair or crutches might affect a game, such as musical chairs. In retrospect, I'd failed to realize that the inclusion of such a topic was revolutionary for a book first published in the 1960s.

In contrast, my parents threw epic parties. Decades later, my friends would recall my Halloween birthday party of 1984. As we descended the basement stairs in darkness, my dad spun his recliner so he was facing us and then illuminated his monster mask with a flashlight. All the girls screamed, and one of my friends even wet her pants.

Rides home from slumber parties revolved around "Chinese fire drills" at stoplights—a dozen shouting girls exiting the side door of the van and running in circles around the van as my dad honked the horn and blinked the lights.

My parents crafted homemade doll beds from orange crates and filled them with homemade mattress sets and sheets. If there was a party, I was sure to receive an invitation.

The nurse paused momentarily noting the family photo on my tray table—me and my husband and son descending in an amusement park ride in mid-motion. We're huddled together with soon-to-be-soaked smiles on our faces.

Her eyes opened wide, and she couldn't help but ask in a puzzled voice, "You have a son?"

It was challenging to share my life history with someone I'd just met. Often, my inability to walk had led others to assume there had never been a time in my life that I was mobile. Though slower than my peers, I'd enjoyed activities such as jumping rope, roller skating, and bike riding in childhood, but by the age of thirty-three, I'd lost the ability to walk and transitioned to a scooter. This in no way matched my pace in life as I cared for my then toddler-age son, toting a diaper bag and juggling bottles of milk and bags of Goldfish crackers like any other parent.

Throughout my life, I had faced the fact that my disease was slowly progressing. This meant increasing weakness for me to navigate but, typically, at intervals of five years or more. I'd recognized the last time I'd step off a curb, venture onto an escalator, rise from the ground unaided, or stride across a patch of ice.

Comments from others, often complete strangers, had served as a barometer of my disease progression. Why couldn't I walk? How did I have sex? Where was my caregiver? Did I need help in the restroom? The increasing impetuousness and lack of personal privacy clung to me like tangled seaweed in shallow water—I could rise from it, but there was something ethereal and binding I could never fully wash from my skin.

Years earlier, I'd ignited this same passion as I read the memoir *Joni: An Unforgettable Story*.[2] I was drawn to the book because I had discovered that the author, Joni Eareckson Tada, beset by a spinal cord injury in her teenage years, had learned to paint by holding a brush with her mouth. She later married and engaged in a public speaking career.

Though Joni's story was uplifting, I was horrified by the passages illuminating her stay in a nursing home in the weeks following her accident. Joni's roommates included teenage girls with muscular dystrophy who had lived in the facility many years before Joni's arrival. Though good-natured and friendly, they perpetually wore pajamas and remained in hospital beds day and night. I couldn't help but compare their tragic confinement to that of veal calves.

How limited the lives of these young girls were from the cornucopia of life experiences open to me. Did they dream of a life beyond the limits of their hospital beds—a chance to experience dating, to excel on a college campus, to share life with another, to become a mother?

For me, the experience of pregnancy was life affirming, though my final trimester would find me *temporarily* bound to a bed. We created a euphemism for my unusual living quarters—my nest—and each day, after my husband dressed for work, he would lay beside me, his cheek pressed to my forehead for several minutes. It was a communion of silence for our new family—my husband breathing in time with me, and my son kicking softly within me.

In Jeremy's absence, I faced challenges. Barely able to heave my body upright, I would slide against the bedroom wall, my face and shoulder smushed into an unusual position, my hands grasping for doorframes for leverage to complete the ten-foot journey from my bed to the bathroom.

I'd queried my health insurance for assistance—could a nurse stop by and offer an arm to aid my walking, perhaps once a day over the noon hour? The reply was ghastly—I could receive help to and from the bathroom but only if I left the comfort of my home and husband and moved into a nursing home facility for the duration of my pregnancy. Even a brief, one-hour visit from a nurse or CNA was considered "custodial care" and was not covered by insurance. The insurance representative queried if I was interested. "I'm thirty," I said and ended the call.

Following the nurse's exit from my hospital room, anxiety-provoking items were brought into my room—a bedside commode, a bedpan, and a gait belt—items I hadn't used except in the case of a brief hospital stay. A trolley track, painted to camouflage with the ceiling tiles, snaked above my head in a path from my bedside to the bathroom.

I wished I could feign ignorance as to the purpose of this seemingly upside-down version of Mr. Rogers's trolley track. With the push of a button, I knew that a sling-type contraption would drop and a patient could be loaded into the sling and moved to nearly any part of the hospital room and restroom, thus saving the strain on health care workers' backs as they maneuvered patients with a handheld remote. I had taken only one ride in the motorized sling while hospitalized. As I hovered in the air like fruit in a produce department, I vowed to myself that I'd never do it again.

Hours earlier, my life was a different story. I rose, showered, and dressed on my own and left my hometown by a quarter to eight in the morning. I drove my wheelchair-accessible van for ninety minutes and arrived at my hairstylist's studio for a cut and blowout.

My mother and I had made plans to enjoy lunch together in the seafood restaurant a few doors down from my hair salon. As I waited for her to arrive, my phone chirped with a message from my friend Ali, in Pakistan: *You are far too busy.*

Momentarily, I bristled at his response. Then I contemplated my schedule: *morning drive, hair appointment, lunch with Mom, exchange shirt at the mall, meet with caterers, drive home, United Methodist Women salad supper with presentation on teapots.*

Maybe Ali did have a point. I was the very definition of overly scheduled. My son's confirmation was the following Sunday. It had previously been postponed nearly a year and a half due to COVID.

From his first attempts to attract my attention last spring, Ali had never been a shrinking violet in my social media. He'd read an interview of me posted months earlier in a muscular dystrophy forum. Facing the same diagnosis, he wanted to connect with me. "Allah has brought you to me," he had insisted. My initial reaction was to ignore Ali's numerous messages or perhaps block his account. I was in over my head, and all logic signaled I should not accept the petitions of someone so persistent.

In time, I thought of my pursuit of Priscilla Lopes-Schliep—how tentative I had been to reach out to her—the fear that I might be rejected and the certainty that my hope resided solely in her. I'd approached Priscilla with the most bizarre claim—that we shared a genetic connection, though she was an elite sprinter and I couldn't raise a pencil over my head. She had every right to send me away, to consider my behavior outlandish. But she took me in and listened to me with an open heart—something I hadn't initially done for Ali.

Though I had corresponded with scores of people with muscular dystrophy over the years, Ali sent me something I'd never received. In an eight-second video clip, he captured his Gower's sign—a series of body movements unique to a person with a hereditary muscle disease. As Ali rose from the ground of his village mosque, he braced his palms against his body to leverage his ascent, then pushed off a windowsill to fully upright himself. In these moments, I connected with my father, nine years after his passing, as if a specter of himself rose from hundreds of still frames to perfectly replicate my dad's Earthly body. Click after click, I replayed the thread like someone watching a phoenix rise from the ashes. Though the logical side of my brain knew I'd fallen for a parlor trick of the senses, my neurons bathed in these images like a healing balm for the soul.

As the minutes ticked by with no sign of my mother, I graciously accepted slices of sourdough bread and refills of my Roy Rogers drink. After an hour had passed, I dialed my sister, Janet, to see if she had news of my mother's whereabouts.

"You understand what day it is," she began.

Thursday, I thought to myself, but then it all came flooding back to me. My brother, Aaron, had passed away on September 2, 2019, and today was the two-year anniversary.

"Mom . . . she just needs extra time," said Janet.

Who forgets the second anniversary of a brother's passing?

I treasured Aaron as my Leap Year Twin. Though both of us were born on November 10, not February 29, I invented the term to capture the special bond I had with Aaron. Surely, there was astrological significance to children born exactly four years apart. We shared the same cornsilk hair and cherub faces and a double dose of our family's hereditary disorder.

My mother arrived minutes later with profuse apologies. We placated our grief with clam chowder and salad, alternating between tears and laughter as we recalled memories of Aaron.

Following our meal, we agreed to meet at a department store three blocks away. As I pulled into the top level of the parking ramp, I felt an odd foreboding feeling. The yellow parking lines blurred, and my chest burned with warmth as I pulled haphazardly into a parking space.

Upon reaching the store's exterior entrance in my scooter, I summoned my mother by phone, and minutes later, she came upon a most perplexing image of me. I was slumped forward against the steering column of my scooter with my arms crossed and my head tucked beneath them like a wounded bird. She described my coloring as ashen. I wanted to nap at her home, but she insisted that she drive me to the nearest emergency room.

Once admitted to triage, the seriousness of my situation became clear. My blood pressure was a staggering 235/115. A plastic band was secured about my wrist; it was official—I wasn't leaving anytime soon.

My brother and sister arrived at the ER and somehow circumnavigated the one patient–one visitor COVID policy. My husband was making the drive from our hometown, ninety minutes away. Though my mother had updated the family on my high blood pressure, none of us were prepared for the doctor's words as he rushed into the room.

"You are experiencing a hypertensive emergency event," he said as he entered my ER room with such explosive energy that the doorknob struck the wall. "There are three levels of concern: tier 1, tier 2, and tier 3. This is a tier 1 event, our most severe category," he explained as he described my triad of symptoms—atrial fibrillation, severely high blood pressure, and rising troponin levels in the blood—an indicator that the heart muscle was actively breaking down, which is a biomarker often associated with a stroke or heart attack.[3]

My siblings and I looked at one another with wide-eyed shock. Though the doctor's words provoked fear, it was the timing of this revelation that filled us with silent dread. We had turned to one another in our grief two years ago on this very day. Was every cell in my body somehow overtaken with this knowledge? Had this led to my medical emergency?

Minutes later, I was transported to the ICU department of the hospital. I needed IV meds to bring my blood pressure back to normal—a drug cocktail that could only be administered under ICU observation.

My husband, Jeremy, arrived in the lobby several floors below but was turned away due to COVID restrictions. I tearfully pleaded into my cell phone that he had to find a way to see me, but it was not possible.

I sought escape as I scrolled through my emails and spied confirmation of the delivery of a FedEx shipment. Like the leader of a diamond heist watching their master plan emerge on closed-circuit television, I reveled in shock and delight. I had slipped undetected into a biomedical supply company, conversed in scientific jargon about my order, and had a laboratory product shipped to the doorstep of my home under the guise of being a biotech firm. It had been a test run—I had placed an order for ten dollars' worth of microscope cover slips and had had it overnighted for forty dollars, proving I could do this. I had taken my first baby steps into the underground world of genetic experimentation and emerged victorious.

In 1985, when I was ten years old, I dreamed the *Back to the Future* DeLorean skidded to a stop in front of my house. A manic Dr. Emmet Brown emerged from the ice-encrusted vehicle clutching an innocuous-looking vial of fluid that he promised would transform my life. He'd traveled from a research lab in the future where scientists were tinkering with my genetic flaw and engineering mice to increase their muscle mass by 40 percent in the span of six weeks.[4] He dismissed the myriad of questions I posed. I simply had to believe, he insisted. What he offered was the tactile certainty of my future rescue. My hands shook with excitement and trepidation as I held the tiny vial and wondered how something so powerful could be captured in something so small—my own custom genie in a bottle.

In the summer of 2021, I staged my Wild West quest for genetic therapy from the privacy of my home office, peering into the blue glow of my laptop screen beaming against the tranquil evening countryside. I'd learned it was possible to take advantage of a little-known loophole in the access to genetic technology; the FDA regulates pharmaceuticals, but they do not regulate laboratory products. This meant that a layperson could access the tools of gene therapy provided they had sufficient knowledge of molecular biology.

As the COVID crisis prompted scientists to steamroll ahead with the burgeoning field of gene therapy—crafting viral-mediated gene therapy, such as the Johnson & Johnson vaccine to combat COVID infection[5]—I couldn't help but wonder why I couldn't have my own renegade gene rescued with the same technology? The gene delivery system had been tested on millions of people throughout the world with few side effects. Didn't the potential gain outweigh the risks?

We experience viral invaders every day, though fortunately, in most cases, viral infection is benign. At times, a foreign bit of DNA is dropped off within our cells like misdirected mail being placed in the wrong post office box. Gene therapy aims to repurpose this mass transit of genetic delivery. Instead of waiting for a lucky break—a virus carrying the correct coding for a gene a patient desperately needs, a twist of fate so unlikely it could take a millennium of evolution to achieve—a harmless cold virus is engineered to deliver the precise sequence of DNA, then packaged Trojan horse style to evade an immunological attack by the body.

Though I had visited many biotech websites frequented by scientists, I hadn't yet made the leap to underground consumers of genetic technology. I'm not sure what I feared—a game show buzzer and flashing lights, the word *Unauthorized* flooding my screen, or DEA agents crashing through the window of my rural Iowa home as I emptied my shopping cart at the checkout? Even the "May I help you?" pop-up startled me on a few occasions.

My greatest obstacle was my resistance to pursue anything ethically murky. I was the type of person who returned to the store with an unpaid can of veggies or refused an extra dollar when offered change. Despite this inertia of honesty, I found the brave new world of genetic self-experimentation infused with a renegade spirit that was hard for me to deny.

The concept of genetic testing on oneself was not a novel topic. The journalist David Epstein broached the topic in his book *The Sports Gene*.[6] He attributes the following quote to H. Lee Sweeney, a physiology professor at the University of Pennsylvania: "One method of delivering transgenes, simply pouring them into the bloodstream, is not necessarily safe, but is simple enough that it could be accomplished by a sharp undergraduate student studying molecular biology."

I had turned to David Epstein for advice on the bioethics and legalities of genetic self-exploration. In turn, he'd queried DEA agents he was working with for a pending story. This is his summary: "A person with a life-threatening disorder can travel out of the country, access medications illegally, and, provided they use the drugs for their own personal use, the DEA would not prosecute," Epstein explained. "It's the moment you sell an illegally obtained pharmaceutical for profit, that's the point that you've violated federal law."

It's a technicality integral to the plot of the film *Dallas Buyers Club*. In the film, Ron Woodruff, a Texas electrician portrayed by Matthew McConaughey, is diagnosed with AIDS in the 1980s. Facing the prognosis of thirty days of survival, Woodruff seeks out underground access to pharmaceuticals theorized to lengthen his life. To help others suffering from AIDS, he sets up a storeroom for illegally obtained medications in a motel room where members of his club can browse the shelves and select what they choose for a nominal monthly fee. The cleverly designed system circumnavigates the DEA's guidelines because a transfer of funds does not take place for a specific pharmaceutical.

A number of renegade scientists, aka biohackers, have attempted to showcase the accessibility of gene therapy, including twenty-eight-year-old Aaron Traywick[7] of Ascendance Biomedical. Traywick injected himself with a custom-made herpes treatment live on Facebook in February 2018 at BDYHAX, a biohacking conference in Austin, Texas. Three months later, he died in a sensory deprivation tank at a spa in Washington, DC. Ketamine, a tranquilizer used in veterinary medicine, was found in his body upon autopsy.

Josiah Zayner (currently, he goes by the name of Jo Zayner), once a NASA scientist who'd grown bored in his previous career, dove into the underground of genetic exploration to satisfy his intellectual curiosity. He founded a company called The ODIN with the intent of placing DIY CRISPR kits into the hands of the general population. He injected himself with a CRISPR construct designed to enhance the development of his muscles and is regarded as the first individual to attempt to edit his own genes using CRISPR technology. His story was profiled in the Netflix limited series *Unnatural Selection*.[8]

My case was different. I wasn't seeking the media spotlight or pushing the boundaries of science to spark debate. I wanted to conquer my genetic disease, not only to preserve my very little remaining muscle, but also to create a better world for a newborn or even a not yet born who shared my genetics. It was a pursuit Zayner supported wholeheartedly.

"You're quite a legend in the underground genetics movement," Zayner told me as I reached out to him in a phone call in 2018. He was referring to my family's underground genetics testing that was profiled on an episode of the podcast *This American Life* in 2016.

"Honestly, I've thought about you nearly every day since hearing your story. Those side-by-side photos of you and Priscilla are unforgettable. I hope that doesn't sound strange."

"No, it's not strange at all," I said. "I would think about this all the time if our situations were reversed."

Zayner had attempted to experiment on his muscles by manipulating the same genetic pathway that had gone awry in my body. Of course, a conversation with a real-life person who could potentially benefit from his genetic tinkering was a scintillating experience.

As I asked the inevitable question as to whether he could help me access gene therapy underground, he had a ready-made reply: "I get this call all the time. Typically, desperate parents contact me. They're willing to travel anywhere and pay any price to help their child. Often, it's a case of a very rare genetic condition with limited options for funding," explained Zayner. He went on to say that he couldn't perform genetic manipulation upon anyone but himself.

"I really wish I could help you—really, I do."

Months into the COVID pandemic, biohackers obtained the genetic coding for the spike protein of SARS-CoV-2 virus and produced a bootlegged vaccine.[9] Long before patients were vaccinated, the rogue citizen scientists inoculated themselves. Titer analysis of their blood indicated they had produced antibodies to the SARS-CoV-2 virus.

I would not be a lone explorer on the trail of medical experimentation; numerous trailblazers had gone before me. Their contributions have revolutionized the role of anesthesia in surgery; pioneered catheterization in cardiac medicine; and eliminated the scourge of plague, typhus, cholera, and yellow fever in many parts of the world. In his book *Who Goes First?* (1998), Dr. Lawrence Altman, physician and *New York Times* columnist, sheds light on the largely unspoken pursuits of physicians behind the scenes: "Self-experimenters are the leaders of medicine—Nobel Prize winners, deans of medical schools, editors of medical journals, and heads of organizations, as well as ordinary doctors" (page 13).

Among the most revered of experimental medicine pioneers is Peruvian medical student Daniel Carrion,[10] who is commemorated by a statue in Lima, Peru, and is the hero of ballads sung by medical students today. In 1885, a curious malady plagued the region. Though the condition had persisted for centuries, it was the most prevalent among railway workers and was called Oroya fever, named after the railway line. Those affected experienced red bumps on their skin and in their mouth, accompanied by fever and severe joint pain. The epidemic nature of the illness suggested it was infectious.

The Peruvian Medical Society set up a prize competition to promote research interest. Twenty-six-year-old Daniel Carrion seemed the perfect candidate. Though dissuaded by friends and professors, Carrion believed the best way to understand the disease was to experience an infection for himself. After he lanced the wart of a fourteen-year-old boy affected with Oroya fever, he attempted to inoculate his arm with the pus on the scalpel. This first attempt failed, and one of his professors stepped in to assist him.

Weeks later, Carrion experienced high fevers, chills, and vomiting as well as pain in his bones and joints. He could no longer eat or drink or maintain his journal of symptoms. Still, he remained lucid. His self-experimentation led to the discovery that the same bacterium carried by sandflies was responsible for both the warts and the resulting Oroya fever. Tragically, Carrion died in the same hospital where he'd originally inoculated himself, but his legacy lives on in the disease named after him, Carrion's disease.

In a case not suitable for the squeamish, Dr. David Pritchard,[11] a researcher at the University of Nottingham, was inspired by his early exposure to aborigines in New Guinea plagued by a tapeworm infection. Pritchard noted those infected with tapeworms faced far fewer allergy symptoms than the tapeworm-free population. Decades later, Dr. Pritchard allowed fifty tapeworms to crawl on his skin and burrow into his body to test their allergy-relieving properties. In 2017, it was reported that the tapeworm produces a protein with anti-inflammatory properties, thus providing support to Dr. Pritchard's hypothesis that tapeworms form a symbiotic relationship with their host.

Perhaps the most brazen of self-experimenters is Dr. John Hunter,[12] a medical doctor on a mission to elucidate the origins of sexually transmitted diseases (STDs). Dr. Hunter hypothesized that the more virulent STD, syphilis, was first characterized by symptoms attributed to gonorrhea. He theorized that syphilis could be more easily treated in this early stage of infection. To confirm the symptoms of a syphilis-only infection, he self-inflicted cuts to his penis and rubbed a pus-laden scalpel from an STD patient over the area. To his dismay, he developed symptoms of both diseases—syphilis and gonorrhea. It was later determined that his patient volunteer was infected with both bacterial infections—a technicality he had not previously considered.

However, the story of Dr. Jonas Salk, the physician credited with the development of the polio vaccine, carries the deepest resonance. In the spring of 1954, as the nation feared the virus that maimed and killed within hours of infection, Dr. Salk proposed a novel theorem—a sample of the killed virus, one no longer able to infect, when injected into the body could "trick" the immune system into producing specialized cells to attack the invader when the polio virus was encountered.

To convince parents to trust this groundbreaking strategy, Dr. Salk[13] injected himself and his two children with the newly developed vaccine, though Dr. Salk has offered conflicting testimony as to whether he tested the vaccine on himself. Following this startling experiment, parents across the country encouraged their children to roll up their sleeves to fight illness, and the polio epidemic was largely contained by these efforts.

Fortunately, the most pressing worries for my health did not materialize. Further testing revealed that I had not suffered from a stroke or a heart attack. The doctors suggested that I have a coronary angiogram to rule out an arterial blockage, but because we neared the Labor Day weekend, it was decided I'd remain in the hospital through the holiday weekend and have the procedure the following Tuesday. Jeremy visited each day, and we passed our time watching *Jack Whitehall: Travels with My Father*, a bingeworthy comedic adventure of a comedian and his father touring various places around the world.

My living quarters felt less like a hospital and more like a five-star hotel, especially because we were offered menus from which to order meals delivered to the room. However, the numerous blood draws required day and night were anything but relaxing. Though my veins were prominent beneath my thin skin, they tended to roll. After numerous jabs of the needle, I felt like a human pincushion.

I'd heard stories about the hospital's famed phlebotomist, a nun, so skilled in the art of drawing blood that many patients slept through her bedside visits. She drew my blood one night without my even feeling the prick of the needle. Afterward, she guided my hand beneath the sheets. My eyes fluttered open just long enough to spy the whoosh of her veil as it swept across my arm. Then she was gone.

Despite the nun's best attempts to come and go unnoticed, I was awake. I considered my options—scrolling through my phone or watching television—but neither appealed. Instead, I thought of Ali. The timing of our text conversations could be tricky because we passed our time at opposite ends of the Earth. As I awoke, he would soon fall asleep and vice versa. Despite our geographic divide, I'd connected to Ali and he to me, as if we'd long ago occupied the same lunar rock that broke apart upon entering the Earth's atmosphere—our souls composed of the same cosmic dust, falling on opposite ends of the globe.

At 4:00 a.m., I called Ali and found him in his familiar place, walking the dirt trails of his family's compound in rural Pakistan, the wind dancing subtly in the sparse greenery. Ali's world filled the sterile solitude of my hospital room like the radiant glow of a firefly in the darkness. I had imagined a far more arid climate, one with no vegetation, but the swaying grasslands revealed gardens of rice and melons. The tall stone walls of the compound housed buffalo, goats, and sheep.

In Ali, I'd found a vessel to channel my frustrations. Nearly twenty-five years had passed since I'd been in Ali's place, struggling to uncover my own family's genetic misspelling—a search that had been stymied for four years in the mid-1990s. For me, it had been a matter of human dignity to precisely define what had gone so haywire in my body, and Ali deserved no less.

Without a causative gene, Ali's quest for answers was a ship run aground. There was no way to accurately diagnose his form of muscular dystrophy without genetic testing. As long as he remained in this medical purgatory, his present and future prognosis remained a mystery, and his hopelessness could not be assuaged by research progress. The fact that I had once charted these same waters and yet could find no safe passage for my friend was maddening.

Though Ali had numerous theories as to how to treat his muscle disease, I knew he had to identify his genetic flaw first. Fortunately, I had an extra prepaid testing kit from Invitae, a genetic information company, and I hoped to send it to Ali. I was greeted by the Kinko's store manager, and when I described my wish to send my parcel to a friend in Pakistan, I was offered a look that signaled my luck may have just run out.

"Pakistan is one of the most difficult places to send packages," he explained. "Let me print off some information for you."

The printer hummed into action, launching four pages of notes detailing regulations regarding the importing of packages to Pakistan. The manager highlighted lines of interest.

"First of all, there's a 68 percent tax on the parcel," he explained.

My brow furrowed as I contemplated this sobering fee.

"The next step is defining the value of the shipment."

How would I define this? The Invitae genetic testing kit I planned to donate to Ali contained a plastic saliva collection tube, a zippered plastic bag, and a return mailer. For most people, the contents offered no intrinsic reward. I estimated the value of the plastic and cardboard as five dollars, at most. However, for one with an undiagnosed muscle disease, such as Ali, this test represented a value of seven hundred American dollars—an amount I had prepaid, which covered my initial consultation with a genetic counselor and the cost to sequence 130 genes implicated in neuromuscular disease. Considering the average yearly wages in Pakistan amounted to seven thousand dollars annually, the kit was nothing short of a windfall. All that was needed to complete the request was a saliva sample from Ali.

In addition to these setbacks, the return trip of Ali's saliva sample to the United States was plagued with additional difficulties—my return mailer only provided return within the United States. Customs paperwork would need to be completed on Ali's end, and the documents would require his signature—a sticking point in our conversations. What I had considered a genius plan—labeling a prepaid saliva collection with my name and birthdate and passing the sample off as my own—didn't sit well with Ali. Allah wouldn't help us if we lied, he'd insisted.

I had developed an entirely different perspective. Every gain I'd made in learning more about my genetic disease had involved some type of deception—to do my family's underground blood draw in 1996 required that phlebotomy supplies be lifted from a hospital and a nurse secretly visit our home; gaining journalist David Epstein's interest began with a wild exaggeration in my email subject line: "Woman with muscular dystrophy, Olympic Medalist—same mutation"; and I'd adopted the lexicon of a research scientist to gain a client rate for Priscilla's genetic testing (the cost for clients was half what was charged to individual patients).

Initially, I was puzzled that Ali placed trust in me. Pakistan is largely separated by gender, especially in rural areas. It is the nation with the world's highest rate of "honor killings,"[14] a practice carried out to restore honor when a family member, most often a female, commits a crime against Islam or refuses a marriage pact. One thousand of the world's five thousand honor killings occur each year in this country.

In my conversations with Ali, I considered a rarely described but potent undercurrent to patriarchal society—the pressure this placed on men as providers and the onslaught of stress that arose when work was elusive or impossible to sustain due to illness or lack of opportunity. Despite my assurances this wasn't the case, Ali was consumed by a daily internal dialogue that his efforts to find work were insufficient—"I'm a bad man . . . I'm a dirty man," he would say to chastise himself, though he spent hours a day studying for job placement exams or traveling to an interview site only to realize four flights of stairs stood between him and employment. It was painful to witness.

As a woman, I was more psychologically shielded from the effect of my muscle disease than Ali was. I had greater latitude in my life pursuits—I could earn a master's degree in creative writing—a lifestyle steeped in imagination and short on promised wages. I was raised with the expectation that I would marry despite my physical disability and freed from the concern of being the sole breadwinner for my family. I could exit the track of a college English instructor to become a mother and maintain the accolades of friends and family. I was painted by a brush dipped in many colors.

The wealth and privilege of my nation masked much of the hardship of my muscle disease. I owned a ramp van—an extravagance equal to the cost of a very modest home—which I could operate independently to get to work and to socialize with others. ADA compliance made elevators and curb cut-ins commonplace. Ten years of employment provided Social Security Disability income even as I continued to work part-time. I was provided health insurance with my husband's employment and hospital coverage through Medicare. Isolation from the full effect of my physical disability and the ease with which I moved about my world left little time to consider the plight of others without these safety nets.

On the other hand, there were many aspects of Ali's life that sparked curiosity in me. Daily meals were a communal event, both in preparation and in dining. As I struggled to shop for groceries, prepare meals for my family, and care for my home, my efforts nearly always came up short. I had a "rule of nines," which I humorously shared with my mother and sister: whatever they could do in five minutes—unload a dishwasher, fold a pile of laundry, clear the kitchen island—would take me forty-five minutes.

I had never truly assumed the role of the "lady of the home." Faced with a domestic task, I often felt the onset of a phantom case of attention deficit disorder (ADD)—I had no idea where to start or what to do next. At times, I reinvigorated the voyeur spirit of my teenage years, a time when I'd explore the homes where I babysat, opening cabinets and perusing cupboards for clues about marriage and family life. Now, even in my own home, I find myself playing the "What's in this drawer?" game. At times I wonder what it would be like to live in a multigenerational family—would I like the extra sets of hands and the ease of completing tasks, or would I feel sidelined and incompetent?

Though Ali was plagued by self-doubt, he was changing the course of my life half a world away. I'd questioned the possibility of creating a custom viral vector for several years, even filling a shopping cart to order before backing out at the last minute from the internet checkout line. I was scared to cross the threshold in which imagination met reality. Having Ali in my life changed everything. Opening my eyes and my heart to another's suffering sparked courage within me.

"How many rupees is this?" Ali asked me as I described the novel gene therapy Zolgensma, used to treat infants with spinal muscle atrophy (SMA), a severe neurological disorder. In its most progressive type, it robs infants of the ability to move, sit, suck, and even breathe.

As I entered the numbers into my calculator, adjusting the value of 74.25 rupees to the US dollar, I gave Ali his answer. A single treatment of Zolgensma cost 157,781,250 rupees, a staggering sum for a country with an average annual income of seven thousand dollars (519,750 rupees). I imagined desperate parents collecting all the rupees they could find from their village—from friends, relatives, and online donors—and I couldn't begin to fathom the weight of these coins—a single rupee weighed 4.85 grams. Collectively, the single injection treatment was equivalent to a ghastly 765,239,062 grams of rupees. That's a weight equivalent to three and three-quarters of the Statue of Liberty!

Ali was a far better candidate for gene therapy than I was. Though he experienced great difficulty in rising from the ground, Ali maintained the ability to walk. Undoubtedly, the Muslim prayers he offered five times a day—a series of vocalizations and prostrate movements—kept his body limber, and the 800-meter walk to his town's mosque served as daily maintenance of his strength. His physical therapy melded with his daily schedule and was unlikely to change. Ali's ability to rise from the ground could be assessed visually to gauge the effect of gene therapy. His blood could be analyzed for the presence of creatine kinase, an enzyme that leaks from damaged muscles. A drop in Ali's creatine kinase level, a benchmark of successful genetic therapy, could be used to quantify an improvement in his strength.

In contrast, I no longer had an elevation of creatine kinase in my blood. I was thrilled to share this news with my neurologist. I assumed my condition had plateaued, but she set me straight: I had no more muscle to lose because I'd already experienced such profound muscle loss from an early age. Additionally, my inability to walk or rise from the ground left me without a visual marker to assess improvement. I was not a likely candidate for gene therapy.

When I'd first explored the web page of a company supplying biological materials custom made to order, I was a ball of nerves. To begin with, I didn't know what title to use with my name. Technically, I held a master's degree. Granted, it was in creative writing and not genetics, but would anyone know the difference? What about the FedEx delivery of my order? I couldn't help imagining a driver entering a town without a single stoplight. *Prairie Valley Genetics?* An imaginary name for an imaginary lab I'd borrowed from a now-defunct school district. How was this supposed to work? The quiet of the main street of my town seemed far more likely to welcome tumbleweed than a bioengineered viral vector.

As I returned home from the hospital, I found a Styrofoam container shipped from a biotech company among the mail scattered on the kitchen island. My heart swelled with happiness. From my hospital bed, I had witnessed the package's launch outside the country and its FedEx travels to my doorstep. Inside the package, I knew I'd find a ten-dollar shipment of microscope cover slips and a forty-dollar charge for shipping. However, this delivery meant so much more. This first shipment had been a test run—a chance to discern if a layperson could breach a biotech supply company, and with Ali's motivation, I had done it!

MARCH 2, 2022

As I closed out my Zoom Gotham Writers Workshop and perused my email, the logo of a biotech supply company caught my eye. *This can't be real. This can't be happening*, I assured myself, but the invoice extinguishes any doubts. Per my instructions, scientists have successfully cloned my faulty gene, SMAD7, into an adeno-associated virus. Five tiny vials, each filled with two hundred milliliters of my personalized gene therapy, have been prepared and stored in a laboratory freezer. My "cure" was now a tangible entity and scheduled to be shipped to me on Monday, March 14. Barring a customs delay, I would soon be holding my miracle worth $2.125 million for the cost of $1,050.

For the time being, I could focus on Ali. I found a place in Karachi willing to test Ali—the Aga Kahn University Hospital. I had vetoed Ali's previous choice—an antiaging clinic without a licensed medical doctor and a Nigerian Herbalife specialist peddling a cure-all vitamin treatment. Ali had sent me the advertisement—a photo of a large woman in a wheelchair and another of a woman walking on a treadmill along with the herbalist doctor's claim that the woman had been cured of limb-girdle muscular dystrophy (LGMD) after a three-month course of treatment. I hadn't wanted to dash Ali's hopes, but the before and after photos were clearly not the same woman.

When he'd first told his parents of his suspicions that he had weak muscles, they were convinced there must be a cure available for Ali. The concept that people in the United States lived with such life-limiting effects involving skeletal musculature seemed an impossibility to them. I explained to Ali many times over that this indeed was the reality for Americans. In recent years, extraordinary treatments involving a one-hour IV infusion of gene therapy had led to substantial gains for babies and toddlers with SMA. Hopefully, this type of technology would be available soon for Ali's specific genetic condition, but we were simply not there yet. First, we had to get Ali's genetic testing results.

Ali understands this reality all too well. Gene therapy is years from development and marketability to LGMD patients. In this span of time, Ali will likely lose his ability to walk. In a third-world country, waiting is not an option.

Ali and Papu arrive for their appointment more than ninety minutes early. The receptionist suggests they explore the clinic grounds in the interim. There, they pose for photos like tourists on a Caribbean getaway—Ali seated on a bench before a palm tree and Papu reclining on a well-manicured lawn.

I take a virtual trip with Ali as he sets off to solve the mysterious genetic illness that has plagued many in his village. As I watched the updates from my home laptop, I took in the sights and sounds of an evening arrival in Karachi, Pakistan.

Three-wheeled rickshaws maneuvered the bustling streets, and decommissioned school buses burst with color as if they'd channeled hundreds of paintball strikes. Motorcycles zipped in and out of traffic with little regard to lane space or right-of-way, passing so closely to vendors that it could be possible to extend an arm and swipe a banana from a peddler's table.

Women are rarely seen in this melee of the busy Pakistani motorway. The many storefronts sparkle in the dimming sunset like the teeth in a generous smile. Men sit in barber's chairs, faces lathered for a shave, and patrons cruise Kirana stores, stocking up on daily necessities. In the moments the traffic stalls, drivers take the opportunity to squeegee the exterior of their rickshaws to a glistening shine or recline placidly against their vehicle in the humid breeze.

Ali Buriro's rickshaw travels at a 25 kilometer per hour clip. He and his friend Papu Hussein have traveled nearly nine hours for a genetic test that could change Ali's life. If Ali has a gene mutation linked to LGMD, as we suspect he does, he faces a severe prognosis. LGMD attacks the large, stabilizing muscles of the pelvic girdle and the shoulder girdle.

Patients with LGMD mutations often lead "normal" lives until a sudden, peculiar event rocks their world. Ali had lived a normal life with muscles of average strength until the age of eighteen. However, he does recall situations in adolescence when he was teased for having a body posture a doctor would describe as lordotic.

It became clear that Ali faced a serious muscle problem as he was playing a game of cricket; as he ran to the catch the ball in the outfield, his legs buckled under him. Ali brushed off the incident to his brother and cousins. Later, his parents realized something was very wrong when he took a fall off a step and crumpled to the ground.

Ali faces difficulty walking the dirt paths of his village—a tenet of his faith requiring five trips to the village mosque each day. A misstep on a pebble can send him careening to the ground. Day by day, Ali can sense his weakness increasing. On one occasion, a cobra appeared on the dirt road before Ali. With no ability to run, Ali explained that he shook with fright and prayed to Allah. Fortunately, the snake did not flare its hood and slithered out of sight moments later.

Lal Khan Buriro, Ali's home village, consists of fifty households on eight acres of land with a population of one thousand. The nearest city, Thul, is four kilometers away. The surrounding villages radiate from Thul like spokes on a wagon wheel.

The Buriro family ancestors emigrated from the Baluchistan province in 1970 to the Sindh province of Pakistan. Ali's grandparents lived a resolutely humble lifestyle, manufacturing household staples from butter to textiles. The relocation to the Sindh province was a vertical move for the family. They traveled with livestock—buffalo, goats, sheep, and horses—and purchased eight acres of land. Family folklore suggests that they each owned only a single outfit of clothing.

In Lal Khan Buriro, there are no hospitals or schools. Electricity is available for the townspeople but, at times, only for half an hour a day. An old hospital in the village center has not been staffed since 2010. Three schools—two for the instruction of boys and one for the instruction of girls—have fallen into disrepair.

"We live a hundred years behind you," Ali explained to me one day as my iRobot scurried about the kitchen. I felt like my life must seem like an offshoot of *The Jetsons* to him.

Later, in the exam room, Dr. Salman Kirmani greets Ali and takes a brief history before examining him. He is pleased to see that Ali navigates the room with ease, but when asked to raise his arms above his head or to stand from a chair, Ali encounters difficulty—both signs of muscular dystrophy. He suspects Ali does have LGMD, the diagnosis provided when he was twenty and first experiencing symptoms. Now, twelve years later, the outlook is the same: there is no treatment for LGMD anywhere in the world.

Like Ali, I have struggled to understand why there have been staggering leaps forward in the field of cancer research but such sluggish progress for muscular dystrophy. "Because cancer strikes down people in the prime of their lives—parents, scholars, heads of industry," my neurologist explained.

"Well, what am I?" I had voiced frustration, thinking of my husband, my son, my life, and my career.

Weeks later, Ali texts me with the good news—his genetic testing results have arrived in his inbox. He shared this pivotal life moment with me, and I did my best to explain the test results. Ali does indeed have a form of muscular dystrophy—calpainopathy. Fortunately, Ali's mutation, in a gene called calpain-3, is the most common type of LGMD. I show him how to access support groups and learn more about websites specific to his gene. Hopefully, in the not too distant future, clinical trials will open for participants like Ali.

10

GOING TO PAKISTAN

Finally, I had a plan. Now, I just had to share it.

I started with my friend Kathy Stearman, a retired FBI legal attaché who had served as the top US representative of the FBI in countries such as India, China, Nepal, Bangladesh, and the Maldives. Of all the people I knew, I figured she would relate to my pioneer spirit, but as I proposed my plans to travel to Pakistan to secure a DNA sample from my friend Ali, I was greeted with an unambiguous no response to my travel plans.

"You never shied away from a difficult circumstance," I said.

"Jill, when I did all the things I did overseas, I had a diplomatic passport and diplomatic immunity. Yes, there were a couple times I could have been killed, and that would have been the end of my story. But all the other times I was out and about, I knew that if I went missing, the US Embassy and the FBI would come looking for me for a number of reasons, including the fact that I, at the time, was a diplomatic representative for the US. The other because I had a head full of classified information.

"You, on the other hand, would be going on a tourist visa. As a woman, and a blonde woman at that, you will be noticed everywhere," explained Kathy. "If you were to go, you would definitely need to be accompanied by a man."

I imagine Kathy struggled to tether my butterfly wings to a male chaperone. She rose in the ranks of the FBI by succeeding in a man's world—completing an obstacle course by maintaining her composure when she was often the only woman in the room. As a consummate feminist, I knew part of her admired my frontier spirit, but even she had stared down darkened alleyways in Mumbai and sought the safety of traveling in a pack.

My friend Joe, who lived in India for the first fifteen years of his life, was also less than pleased with my travel plans, particularly my wish to take him as my male chaperone: "Our mothers would meet for the first time as they were arranging for our dead bodies to be returned from overseas."

I was used to Joe's playful banter, and I sailed past this predicted demise.

"We can catch a direct flight from Houston to Karachi," I gushed with excitement.

"You know I'd go anywhere to help you, but we are not going to Pakistan," he said definitively.

Joe was going through a difficult divorce and had turned to alcohol to numb the pain. He rarely left his home, where he worked as a web designer. This trip to Pakistan was my not-so-subtle-ploy to launch Joe past his agoraphobic tendencies.

"You could translate for us," I suggested.

"Where were you in world history class? Indians and Pakistanis are enemies."

"Why?"

"Do you want the long answer or the five-minute synopsis?"

"Five seconds."

"Indians. Pakistanis. We all come from the same race of people, but there was so much turmoil that the British drew a line in 1948—Muslims to the north, Hindus to the south."

"You could just blend in . . . ," I began.

"I am dark . . . Southern India, and beyond that, I'm now living in the United States. They can smell the West on me."

"But . . . ," I stammered.

"Look, you think you are going to get off the plane, and it will look like a melting pot, like in the USA. But instead, you'll see a sea of brown faces. No one looks like you. Everyone is looking at you." Joe's speech slowed for emphasis. "Now, *you* are a target."

As I circled back to Kathy, she mirrored the same sentiment. This kind of travel could be done, but she insisted travel overseas required much advance planning and the hiring of private drivers, security detail, and so on.

"I think you need to consider the fact that Pakistan is much like India in its infrastructure. It would be difficult for you to move around, and I'm sure the streets and sidewalks and entranceways are every bit as broken and in disrepair in Pakistan as in India. Let's not forget the food and toilet facilities," she continued with an air of confidence as only one who has experienced this firsthand can convey. "I put those two together because the consumption of one definitely leads to the other in that area of the world."

My romantic notion of world travel began to fray, and I no longer envisioned myself as the intrepid novelist finding inspiration in a faraway land. I couldn't help imagining my literary agent's concern as I updated my location via an unreliable wireless connection: "You're where?"

"I know that your heart is in the right place, but I would caution against making this trip," said Kathy. "I'm not trying to talk you out of it. I know you are every bit as strong-willed and stubborn as I am. I just want you to consider it very hard."

I felt like a cornered steer charging toward a closing pasture gate. As my muscles have weakened, I couldn't help but consider the effect this would have on my ability to seek out life's adventures. Of greatest concern, I would likely soon receive an ICD, an implantable cardioverter defibrillator, a device used to regulate my heart's electrical signals and provide a shock if I experienced a dangerous arrythmia. Beyond the obvious fear of being shocked, I was concerned about the several weeks of recovery, a time when I would not be able to use my muscles to perform daily tasks, such as transferring. Would I lose independence with this procedure? Kathy was offering heartfelt advice, but much of me wished I could have lived the adventure she had experienced in her FBI career had I been born with a healthy body.

"Pakistan's a tougher nut to crack than India," she suggested as if she could read my thoughts.

Joe's final response left no ambiguity.

"If you ever think about going alone to Pakistan, I will drive from Houston to Des Moines. And I will track down your husband and all your family and friends, and I will say, "You have to stop her."

The musings in my mind stalled as Joe reeled me back to reality. In the past eighteen months, he had filled his gas tank only twice because he worked from home. He sustained his social connections with Uber Eats deliveries and cared for a family of feral kittens. This pledge to drive hundreds of miles was testament to how much he cared.

"I'd like to think I'm a reasonable guy and not some misogynist jerk," my husband, Jeremy, responded as he leaned closer, inhaling deeply, "but there are a few things I forbid my wife to do." His face reddened as his fingers combed through his hair.

"In fact, I'd like to think it's a very short list," Jeremy continued. "But my wife—flying solo to Pakistan?"

"Qatar Airways Flight QR726."

"You have a ticket?" he asked with incredulity, "without even telling me?"

"Well, I can cancel it," I suggested.

"Oh, no you can't," he insisted, "not without a hefty fine."

I was frozen in place like a child clutching a teddy bear with a chewed ear and a going-to-Grandma's suitcase who had just announced her plans to run away.

"Well, it looks like you're going to Pakistan!" Jeremy offered as a stiff rebuke. "Are you all packed?"

My mouth grew dry. The words chastised my tongue in place. Did this mean I shouldn't complete my visa application, I wondered, but thought better than to ask. I recalled the difficulties in attempting to attach a suitable facial photograph to my tourist visa application—*Can't detect eyes . . . Can't detect face . . .* I wasn't successful despite dozens of tries. Was the universe trying to tell me something?

"When did you make this reservation?" Jeremy asked.

"I made it when Amanda died."

Our friend Amanda, who shared my diagnosis of Emery-Dreifuss muscular dystrophy, died at the age of thirty-six, leaving behind a thirteen-year-old daughter. Jeremy rose from his recliner and stepped toward Martin's room to ensure his door was closed. As he returned to our bedroom, he closed our door too.

"I'm afraid I'm going to die," I sputtered.

I've made the perfect strike at the *imaginary* piñata that hung over our lives. A sprinkling of pretend foiled candies spilled from the hoof of a papier-mâché horse. Over our seventeen-year marriage, Jeremy's blindfolded swings have come close, but we've both shied away from reality talk. As he returned to his chair, eyes locked with mine, his facial expression sobered.

"Sometimes I'm afraid you're going to die too," he said for the first time.

Jeremy's swing severed the imaginary piñata with this strike, and its entrails poured candy on me like a waterfall.

"Do you understand why I have been doing the Nutrisystem plan?"

It seemed an odd question. I hadn't thought much about the diet plan shipped to our house each month. I shrugged my shoulders.

"I want to be able to take care of you," he explained. "I'm not as strong as I was twenty years ago. I can't just toss your scooter in the back of my truck."

My body tensed with resistance. There were things I feared more than death. The concept of not being able to care for the most personal aspects of my life was a topic I wanted to delete from my brain's motherboard.

"Look, I've got a brother and sister I'm not particularly close to, a father who's out of the picture, cousins I almost never see, and an uncle who calls a few times a year," he paused for effect. "You and Martin are my whole world."

I recalled our first weeks of dating when Jeremy made a heartfelt observation. Previously, as a single man, he often went an entire weekend without saying even one word aloud. It was such a contrast to the way I lived my life. I was known to strike up a conversation with the caller of a misdialed number, banter with strangers in a bookstore, or chat freely with the checkout clerk at the grocery store. Isolation was something I'd rarely experienced.

"Martin's going away for college in a couple years," he continued. "I don't want to be alone."

His admission struck a chord with me. Though I was very happy in our marriage, I encountered bouts of aloneness, an emotion I experienced that was so different than loneliness. Exhausted from battling symptoms of my genetic disorder over the past forty-five years, I feared what the future held for me as my muscle weakness slowly progressed. I longed for connection with others who had faced similar odds.

I had grown self-absorbed in recent years, seeking external praise from others to lift my spirits. My greatest fear was to find myself strapped to a gigantic power chair, clad in sweatpants and Velcro shoes, my unkempt hair overdue for a wash and blow dry.

"I need Ali's story for my book," I countered, returning my wandering mind to the discussion at hand.

I had recently finished reading *The Naked Don't Fear the Water*, an account of a journalist's escape from Afghanistan. The harrowing tale described the underground journey as the author cast off his Canadian passport and assimilated as a native, alongside his friend, a former translator for the US government. Together, they faced many hardships—crowded refugee camps plagued with poor sanitation, bounty hunters seeking fortune, and a treacherous passage in the turbulent Mediterranean waters in a flimsy raft and less-than-seaworthy life jackets.

My preference was to travel with Jeremy, especially when airplanes were involved. Nothing felt better than the bounce of his step as we descended an airplane ramp piggyback style. As he placed me effortlessly in my assigned seat, the relief of the flight attendants was palpable as they rolled the aisle chair sans rider off the plane.

I had convinced myself I could handle a solo trip halfway around the world, even though I had not been able to stand or walk in fourteen years. There were two things I could do with ease, sit and wait patiently, and a plane trip halfway around the world required an abundance of both. Of course, this mission wasn't without obstacles. Travel without Jeremy required the ordeal of being strapped in the "Hannibal Lector" chair, as I'd named it—arms crossed over my chest, my body was secured to an aisle wheelchair by numerous seat belts and then lifted across the threshold of the plane entrance by flight attendants. Once embarked, I would be rolled down the aisle, often in the presence of a full cabin of passengers. It was a humiliation I was willing to endure to save Ali.

I assumed my proposed layovers in Chicago and Doha, Qatar, would provide ample time for rejuvenation. For airport layovers, I was often assigned residence in "special needs"—an expansive room with a big-screen TV, comfortable chairs, cots for rest, and a concierge to address my needs, from fast-food cravings to cell phone charging. At least, this is what I'd come to anticipate from my American leg of the journey.

"You already have plenty of material to write your book," said Jeremy. His words ushered me back to reality.

"No, I don't."

"You know what I think?" Jeremy prodded. "I think you're afraid to fail. You're afraid to put your story out there just as it is right now."

I found some truth in Jeremy's words. I had moved the goalposts of my story innumerable times. I had let the rejection of previous queries to literary agents hinder my progress.

The stakes of a medical memoir were so high. Readers wanted the satisfying stand from a wheelchair or a robust legal win against the pharmaceutical industry in David and Goliath fashion. Where was the arc of my story if I made no such leap?

Setting out to help another person was how I found the strength to continue, to find purpose in my life journey, even as my body slowly weakened year by year. I wanted to travel to the other side of the world to find stories that had never been told, such as that of Ali's cousin, Jan, a mother of five who had learned to live homebound as her muscle disease robbed her of her ability to take on the outside world. I wanted to sit by her bedside and capture her words, words that would not reach beyond her tiny village without my help. I wanted to map the genetic disorder plaguing Ali's family by constructing a family tree and taking DNA samples, all the while collecting family stories, bringing life to the branches like newborn spring. To solve a medical mystery, to chart a new horizon of hope, was like cocaine to my brain.

"I want to write an incredible story. I want to travel to the other side of the world and talk to people who need help," I insisted. "I want to be like Clarrisa Ward on CNN interviewing mothers and children in Afghanistan."

"I have no doubt that without your disease, that's exactly where you'd be," said Jeremy with conviction, "but you're not!"

I bristled in frustration. I wanted to release the angst in me like a bubbling beaker. It was surprising, even to me, that the passage of more than forty years with a significant disease hadn't lessened my desire to be an immersive journalist. If anything, it had stoked the fire. The more I couldn't do with my physical body, the more my mind strengthened.

"Do you understand what could happen to you? You could be thrown into the back of a van and end up part of the sex trade in a foreign country. I might never see you again."

I pictured myself a passenger in a manual wheelchair, the most helpless of medical devices. Facing forward, I would not have the opportunity to speak to the person pushing me. I would not have a way to change the direction of the wheelchair or stop its forward momentum. The identity of my assistant could change midstride without my knowledge. I could be pushed into a windowless room or out a back entrance without my consent. The harsh warning sunk in.

"Pakistan doesn't exactly have a good record on women's rights," Jeremy insisted.

I considered the clothes I'd planned to wear on my journey to the Jinnah International Airport in Karachi, Pakistan—a button-down shirt and dark denim jeans and a scarf to cover my hair. For the first time, I considered the danger I faced. It was not about what I considered modest clothing; it was about what Pakistan considered modest. I couldn't begin to understand the cultural norms of a distant country. I was in over my head.

"Why do you have to be the savior of Ali?" Jeremy protested.

The question silenced me. In the year that I had been talking to Ali, he had morphed into the physical embodiment of my dad and brother. I had not been able to save them, but maybe I could save Ali. Once I had accepted the mission, it was nearly impossible for me to contemplate the potential consequences of my actions with clarity.

"No one is ever going to show up in his village and say, 'I'm here to help Ali,'" I explained. "No one will but me."

The stakes for Ali were so much higher than the stakes for me. If Ali did indeed have a mutation in a gene resulting in a limb-girdle muscular dystrophy (LGMD), he faced a severe prognosis. As the diagnostic name implies, LGMD attacks the large, stabilizing muscles of the pelvic girdle and the shoulder girdle. Without treatment, Ali faced tremendous disability, particularly in a small village traversed by dirt paths, miles away from basic infrastructure.

Patients born with LGMD mutations often lead "normal" lives until a sudden and peculiar event rocks their world. My friend Angie, who has a diagnosis of LGMD, lived a normal childhood until a freakish dismantling of her run occurred during a softball game at the age of eight. She had had a clean run from home to first base, but as she attempted to steal second, her legs collapsed beneath her. She fell abruptly, kicking up a plume of dust. In the stands, her mother rose to her feet, shielding her eyes from the sun with her palm. The two—mother and daughter—locked eyes, each silently contemplating how their world had changed in mere seconds.

The sudden onset of muscle weakness is a common story for those affected by LGMD. I knew of a farmer in his sixties who recalled slipping on husks of corn that littered his field, something he'd navigated with ease for decades. Another farmer, slightly younger at his age of onset of weakness, described reaching his hand to the top of his tractor to pull himself into the cab and his hand's slipping with each attempt.

Bliss, a friend of mine as sweet as her name, first noticed signs of LGMD as a teenager in her step aerobics class. Her once-nimble legs failed to negotiate her step platform, and she fell abruptly, startling herself and her instructor. After many weeks of testing to rule out more common disorders, such as multiple sclerosis, it was found that Bliss had LGMD.

Like many others with LGMD, Ali had lived a normal life with muscles of normal strength until the age of eighteen. Ali's first recognition that something was wrong was eerily similar to Angie's story. He had been playing a game of cricket, and as he ran to catch a ball in the outfield, his legs buckled under him. He brushed off the incident to his brothers and friends. Later, when he took a fall off the front step of the family home, his parents realized something was very wrong.

Just days before, my plan had seemed ingenious. My brief itinerary focused on the mission of my visit—to collect Ali's saliva for DNA testing and visit with him for a few moments to authenticate the "real-life Ali" from the long-distance one in my book. As I researched flight plans, I devised a quick exit strategy—I could arrive in Pakistan a little past 7:00 a.m. and catch the departure flight back to the United States three hours later.

Chapter 10

There was nothing untoward about my meeting with Ali. This early-morning visit would stay within the bounds of the airport, a public place. I'd likely be leaving on the same plane I'd arrived on, so odds were that the plane wouldn't be grounded with a mechanical delay. Even with my attention to detail, the humor of *Gilligan Island's* "three-hour tour" hadn't escaped me.

The reality of traveling into Pakistan had dimmed for me as I read warnings from the US State Department urging Americans to rethink travel to Pakistan. The ousting of Prime Minister Iman Khan had prompted uprisings in Islamabad and Lahore. I had visions of descending into Karachi as thousands took to the streets in protest. When I'd booked my ticket, I didn't have a full understanding of the dangers of the visit, even being limited to just the airport. I would be bound by the laws and customs of Pakistan from my moment of arrival.

Later, I dialed Qatar Airways, and the relaxing notes of a zither serenaded me. The hold lasted a mere ten seconds, then a male voice answered, smooth as rose petals gliding on water. I felt like a humble beggar rescinding my wish granted by my genie in a bottle.

"Once I cancel this ticket, it can't be created in the system," the representative explained.

Rivulets of tears ran down my face. This wasn't a leisure trip I was planning. I was seeking the "just ten minutes more" with a loved one who had crossed over to the other side after death. In the year since I had been talking to Ali, my pangs of sadness were assuaged by the ping of my cell phone receiving a text message. Canceling my trip to Pakistan was like asking me to spy my dad's gait among a crowd of people and subdue the urge to hug him and gush with updates of the nearly ten years we'd been apart.

My yearlong quest to secure genetic testing for Ali was also derailed by this change of plan. I wouldn't have the chance to swab the inside of Ali's cheek for a saliva sample and tuck the Invitae test kit in my carry-on bag, secure in the promise that I would have Ali's gene of interest identified in two to three weeks. I wouldn't be able to search the literature for the promise of gene therapy trials specific to Ali. I wouldn't be able to define the precise type of muscular dystrophy affecting Ali and members of his extended family. This was my one chance to play the role of international spy for just a few hours, and my mission had gone bust.

"Once again, Miss, I need to remind you that this ticket cannot be replicated once canceled."

Couldn't there be some type of hassle? Maybe a nonrefundable cancellation? But, no, full reimbursement for the canceled trip would be returned to my credit card within seven to ten business days.

As the dates of my itinerary passed, I searched the news for information on my clandestine travel plans halfway around the world. The day before my intended arrival, a female suicide bomber detonated a bomb, killing herself, three Chinese teachers, and a bus driver outside Karachi University. Temperatures swelled to 129 degrees Fahrenheit, and the observance of Ramadan forbade the consumption of food or drink in daytime hours. I reached out to my sister, Janet, on the day I would have arrived in Karachi.

"Is it wrong for me to feel sad?"

"Jill, if there's one thing I can say to you with certainty, it's that this trip would have gone nowhere near the way you thought it would go in your head. We're not talking about the Des Moines Airport," she insisted.

I considered her words as I recalled Iowa's largest airport—two concourses, labeled A and C; piped-in Muzak; and scrolling advertisements of blonde pigtailed children chomping down on sweet corn at the Iowa State Fair.

Her words sank deeply. I knew she was right.

11

BREACHING THE IVORY TOWER OF MEDICINE

It was the type of Christmas card salutation that prompted me to pause in the post office lobby:

"Was sick for a while
But getting better
2022 will be better"
Tom & Mai & Therese & Tomy

No photo was included.

The card was simple. A dove with an olive branch offered hope for the new season. It was a sparse message—unassuming, reserved, solemn, but suggesting of so much more. I'd been sequestered with my family for nearly a year and a half, and we were finally making brief, calculated appearances in the outside world as the threat of COVID waned. What had become of friends and family fueled my curiosity.

Nearly thirty years had passed since I'd seen Dr. Tom Horiagon in person. I was an intern in his laboratory in the summer of 1994, and we published a paper together. It was a remarkable climb for a nineteen-year-old, and I'd checked in with him over the years with updates in my life—an internship at Johns Hopkins, graduate studies in creative writing, marriage, a son, a book deal.

We'd corresponded over the years, but more often than not, I spun through Dr. Horiagon's life like a whirling dervish, eager to test out theories of medical advances that I had hoped would someday transform the course of my genetic disease. Time and time again, he played the sounding board, allowing me to channel my frenetic energy until I wound down, exhausted in my efforts, once again assured with the knowledge that conquering my disease held no easy answers.

I sent a brief email but received no response until June:

June 18, 2022

Hello Jill,

It is great to hear from you.

I have been facing a health struggle of late. I developed painful tongue ulcers out of the blue and it turned out to be a fairly aggressive cancer. I don't have risk factors for this, but it happened anyway.

I underwent pretty aggressive treatment with surgery, radiation, and a little chemo. I had to stop almost all work because it was too much for me to remain very functional. The tumor seems to be gone, but I developed problems from the radiation and underwent two more procedures to cover up exposed mandible. The last procedure was just a month ago and I am healing pretty well. The right side of my jaw is swollen and will resolve slowly, so I am growing a beard to be presentable. I lost weight with all of this, but I am slowly regaining it. My speech is functional, but it sounds foreign to me.

I haven't really been sick before, and I'm not very good at it. These days, I feel mostly injured post-operatively, and I don't feel systemically ill. I am frustrated that this is taking so long to get through. The COVID isolation and the general "strange days" feel of modern times has not helped much for that matter.

I haven't been as available to you as I should be. Some of this was about not wanting to describe being a patient to someone as courageous as you. Some of this was also about moments when I didn't feel particularly triumphant and gave into some self-pity.

I will do better.

I'd arrived for the first day of my internship in 1994, fresh-faced and full of optimism despite the turbulent storm brewing in my mind. My neurosis extended far beyond the reach of "oldest daughter syndrome." I was at the helm of a charter boat filled with three younger siblings and a father affected by genetic disease, not to mention, I also had a firsthand immersive experience with our rare form of muscular dystrophy. There was no treatment, no answers, and frankly, no hope, and that's why I'd stepped in.

I was led into Dr. Horiagon's office at the Human Gene Therapy Research Institute, a biotech start-up focused on gene therapy for cancer that sprouted beneath the lowest level of a parking garage in downtown Des Moines. Most people entering and exiting the bustling hospital and clinics didn't know of its existence. My genetics professor at Drake University had set up an internship for me with Dr. Horiagon.

As I waited in his office, my eyes scanned the walls, taking in the many diplomas from Brown—undergraduate and medical school—as well as degrees from the University of Michigan and UC San Francisco. Dr. Horiagon arrived momentarily, and after the pleasantries of introductions, we attempted to find a laboratory coat to fit me from several preselected styles. This wouldn't be an easy task—I was barely eighty-seven pounds soaking wet. I felt like a child playing dress-up because my arms were as animated as walrus flippers in the dense fabric of one of Dr. Horiagon's lab coats.

Later that evening, I visited a uniform store to help me with my too-big laboratory coat problem. Two Asian ladies waved disinterestedly from the back of the store, both peering inches from the black-and-white television set as O.J. Simpson and Al Cowlings made their low-speed escape from authorities. I found a jacket to fit me well and selected cornflower blue thread to stitch my first name in cursive on the lapel. It was official.

The following morning, we exchanged laughs as it became clear that the slim-fitting jacket I had personally embroidered was, in fact, a phlebotomy jacket. No matter. It served its purpose, and it explained the numerous pockets.

Dr. Horiagon had crafted a novel laboratory project to teach me the basics of genetics while also ensuring the study held global resonance. I would be creating a one-of-a-kind deletion map of a phosphorescent gene in the jellyfish *Aequorea victoria*. In simple terms, like a firefly uses the luciferase gene to glow in the dark, this breed of jellyfish uses green fluorescent protein (GFP) to glow in the depths of ocean waters, an extraordinary testament to the beauty of nature's bounty and a discovery that provided scientists a reliable biomarker for experiments.

My summer project involved removing amino acids, one by one, until the glowing properties of the GFP were extinguished. The resulting mutation map would allow scientists to know how much of the GFP gene could be removed and the resulting protein still glow.[1] This was important because scientists were often short on space in designing gene therapy vectors, and anywhere they could trim back contents would allow them to transfer larger genes of interest.

To cement the utility of the project for me, Dr. Horiagon and I visited the laboratory darkroom. After we'd donned protective UV shields, he waved a UV wand over two petri dishes filled with bacterial colonies. The dish on the left remained dim, but the one on the right glowed green in about 20 percent of the colonies. He asked me if I'd rather select the colonies that had successfully incorporated GFP by guessing or if I'd prefer to use GFP as a reporter gene. Of course, I chose the reporter gene, and the warm glow with UV illumination told me exactly which colonies to select.

In a laboratory, GFP serves as a tag, often attached to another gene, providing quick reference as to whether a gene of interest has made it to its desired destination. Open the deep freeze of a laboratory in nearly any country throughout the world, slide a cell sample beneath a microscope lens, and chances are you'll uncover the warm glow guiding you to cancer cells responding to a novel treatment or a swarm of bait fish testing out bioluminescence in a UV-illuminated aquarium.

By summer's end, my laboratory skills were vastly improved. I could "race" samples of negatively charged DNA with electricity, move a gene from one organism to another, and select an enzyme to snip DNA like scissors to paper. Somewhere along the way, I'd purged my status as the lowest member on the totem pole of genetic destiny and emerged as the Eliza Doolittle of molecular biology, morphing my Iowa nice to Providence confidence with glacial velocity—but not without its setbacks.

"Can I talk to you?" I asked.

"Sure," Dr. Horiagon said, sliding his maps and pamphlets of future kayak expeditions from the lab bench to a drawer.

"No, in your office," I clarified.

I'd promised myself I wouldn't cry. Crying had been the entry point to my internship, but I had to pull it together. I was here for a purpose. My future depended on understanding the powerful and intimidating field of genetics.

"So, if my disease is dominant, then how are things supposed to be if I want a family? How do I talk about what happened to me, what's going to happen to me?"

The translation: How do I go on when I feel like I've struck out—at nineteen?

The look on Dr. Horiagon's face mirrored the gravity of stepping off the marked hiking trail of medical consultations, sinking into spongy peat moss, the blurred passage for one. A conversation of this gravity had been drilled out of him since semester one of medical school. Physicians needed to maintain a healthy emotional distance from patients. He was entering a void where he'd received no training, for which he had no answers for the pleading eyes of a teenager restrained in the starting blocks of life.

Doctors occupied designated spheres of space and time, populated with team members—nurses, lab techs, social workers—to keep the medical hive humming. Most importantly, the preestablished machinations of a clinic schedule were meant to keep a patient like me at bay—the kind that overshares and underlistens and siphons successive twenty-minute consultations like a hungry person devouring a sleeve of Girl Scout cookies.

My pleading eyes focused solely on him. He had to answer on behalf of all of science for all of the world's failings. In one spiteful reply he could sever it all with a fiery comeback: "I'm not God." Instead, he remained transfixed, refusing to abdicate.

In our conversations, Dr. Horiagon spoke of fractured family elements—paternalism, orphan disease, for-profit hospitals replacing benevolent institutions. I bristled as he described the era of the early 1980s when he studied medicine. Medical students were trained to keep the details of a diagnosis away from the patient. The burden of navigating these waters fell to the physician; a patient shouldn't be burdened with such minutiae—they had enough to deal with.

"Why would I possibly want to bury my head in the sand?" I asked with irritation.

The concept that the doctors might not have focused on establishing a specific diagnosis for me or my family despite years of clinic visits because they didn't want to trouble me was a startling revelation.

"I'm not saying I agree with this," explained Dr. Horiagon. "I'm just trying to point out how things operated in the day."

"But what about now?" I asked. "I need to find my gene . . . to get on with my life."

"You have what's called an orphan disease. When there's a disease affecting very few people, such as 1 in 100,000, there's no incentive for a laboratory to investigate a therapy for such a rare disease. They wouldn't have the pool of patients to market to in order to recoup research and development costs."

JULY 2022

With Martin off at summer camp and concern about Dr. Horiagon's diagnosis growing, I convinced Jeremy to take a trip out West to visit the doctor who meant so much to me. We selected the lobby of our hotel, the Hilton Garden Inn in Highlands Ranch, Colorado, as a perfect venue for conversation. As we sat close to the entrance, Dr. Horiagon bypassed us but turned swiftly with a warm smile.

"Jill, you look just wonderful!" he said.

I placed high importance on the response to my appearance from someone who hadn't seen me in a while. I knew it was shallow, but I groomed myself for such occasions, confident that these unrehearsed responses served as a barometer of my health.

Dr. Horiagon had shed a few pounds from treatment but otherwise appeared as his gregarious self.

"I know, I know . . . I sound like a Furby," he explained.

It took me a moment to place the reference, but I realized he was talking about a faddish and furry toy from the late 1990s with large round eyes and ears and a tiny bird's beak. The toys represented a first for a domestic robot and were trained to start out speaking the novel language of "Ferby" but gradually transition to English (or one of eight other programmable languages). I loved his association with a novel toy built on continuing education, a reinvention of the self.

I came to understand Dr. Horiagon's speech despite its higher octave. Nearly a third of his tongue had been removed due to cancerous lesions, and he had to relearn to speak and eat and swallow and various other functions no one thinks of unless challenged in this way. We had spoken on the phone to set up plans to meet. I had to concentrate, but I could make out most of his words. His most challenging symptom was a trickle of saliva that ran from the corner of his mouth. Every minute or so, he audibly sucked from the corner of his mouth to draw the saliva back in.

I'd imagine many people would withdraw from society if faced with these issues. The fact that Dr. Horiagon had pursued the study of law despite the assault on his body came as no surprise to me. He had a love of learning that couldn't be extinguished by a health scare. Though most of his classes were over Zoom, he traveled for on-campus classes in New Hampshire a few times a year.

Curiosity questions had decreased in frequency ever since I'd transitioned to a mobility scooter, yet I still bristled when asked a personal question about my body and how it functioned differently. The thought of Dr. Horiagon adjusting to his Furby voice and being hassled for it saddened me. He was, perhaps, the most intelligent person I'd met in my lifetime, and I was sobered by the obstacles he faced because I knew, firsthand, just how challenging this could be.

After moving through a conversation of three decades of life events that brought us to this day, we focused on COVID. Unsurprisingly, Dr. Horiagon ran toward New York City at a time when most people were running away or sequestering. As a pulmonologist by specialty, he was desperately needed. He described scenes straight out of a horror novel—the streets emptied of cars and pedestrians, the morgue over its capacity with bodies, a gurney pushed into an alley and abandoned with a deceased elderly patient with no identity.

"What scared me the most was the look of terror in the eyes of the residents and interns. They clung to the lapels of my coat and looked me squarely in the eye and said, 'Tell me! Tell me! Is this the end of the world?' Many years ago, I was that resident, turning toward Dr. Fauci with desperate fear in my eyes, most likely with questions he couldn't answer for me."

Dr. Horiagon's mission of mercy was cut short after three weeks, but it wasn't the hazards of COVID that drew him away from his post. He'd encountered profiteering by physicians in the wake of the crisis. This, he explained, was an unforgivable blemish on the profession. It was at this point that he left the field of medicine for good and decided to pursue a degree in law.

"I suppose we should move on to the more interesting topic of the visit," suggested Dr. Horiagon.

"Yes," I said with a gleam in my eye. I had shared my exciting update in an email weeks earlier. Dr. Horiagon's response had been captured in three letters—W-O-W.

I'll never forget the excitement when I discovered that my customized gene therapy would be arriving as an overnight shipment in mid-March 2022. I was provided a tracking number and a FedEx delivery schedule, and I followed along throughout the afternoon and evening, noting the checkmarks of success as my gene therapy moved through inspection checkpoints and was loaded onto barges and flights. I was sure something would delay my precious package, but the voyage moved seamlessly without a hitch.

As I watched from the windows of my living room, a FedEx truck pulled up slowly in front of my house. The driver cataloged a few items with a scanner and then exited the truck clutching my Styrofoam box. I circled through my house and exited the garage, certain the package would require a signature, but as I rounded the side of my house, I saw the truck moving down the street. A large Styrofoam container, approximately three cubic feet in volume, rested on my front porch.

In a brief and shaky video, I angled my phone to my face and shared my first impressions upon receiving my treasured shipment. I woke up my son, Martin, from his late-morning slumber with the enticement that I had a secret to share.

"You're pregnant?" he asked as he entered the kitchen.

"No, definitely not that, Martin," I assured him. "Will you help me open this?"

I had attempted to open the box on my own, but I needed more grip strength to tug the packaging tape. Martin removed the tape with ease and opened the Styrofoam lid and peered into the contents with a befuddled look on his face. The box was filled with a few hundred pellets of dry ice. We alternated quick dives into the frozen space, with Martin finally surfacing with a small box in hand. Inside the box were five tiny vials, each the volume of a half-used piece of chalk, and at the bottom of each vial, two hundred milliliters of liquid had been spun down to form a frozen pellet.

I had wondered if I should tell Martin exactly what he'd just opened or provide an ambiguous answer under the circumstances. The answer came most assuredly to me. Martin had had to grow up faster because I was his mother. Though he was just sixteen at the time, he seemed wise beyond his years and more of a confidant than a dependent to me. Besides this, I was curious as to his uncensored response. I was at a time in life that I wanted to be able to turn to him for his rationale about life's most difficult choices.

Ultimately, I told Martin about the gene therapy.

"It's interesting," he said without embellishment.

Dr. Horiagon's reaction was humorous.

"So it's in your freezer between the frozen peas and ground beef?"

We couldn't all help but break into laughter. The levity of his response put a hilarious spin on a complex problem.

"That was one heavy lift," he chimed in.

"So now what?" I began.

"Do I think you would die if you took this therapy?" Dr. Horiagon contemplated aloud. "No, I really don't think so."

These words were a relief to me, considering that my husband was taking in the doctor's advice at the same time I was.

"I could ask, 'What would I do if I was in your shoes?'"

He contemplated this for a few moments. His answer was plain as the smile on his face.

"I think we all know what I would choose, so perhaps I'm not the best person to pose this question to."

This candid response softened this mind-numbing ethical question. It struck me that this type of moral imperative was being played out in various modes of commerce throughout the world. Bioethicists raised concerns regarding how race, wealth, and power shaped which patients would be offered these groundbreaking genetic therapies—the most expensive crossing a seemingly unfathomable $4.25 million in 2024. Hedge fund managers utilized news of patient deaths to dictate which biotech start-ups would soar with the news of FDA approval and which stocks would make a nosedive in the face of patient deaths or adverse injuries. State-run Medicaid programs pondered how multimillion-dollar gene therapy price tags for individual therapy could be siphoned from a state budget meant to provide care for the neediest and most vulnerable citizens of the *entire* state.

"You are right in the concerns you've raised about the cost of gene therapy and the observation this is not sustainable," said Dr. Horiagon.

"So couldn't gene therapy be marketed globally, the cost spread out over many more patients to reach a reasonable price per person?" I asked.

"Ah, but you're forgetting the Pareto principle," he cautioned.

I had spent so much time studying genetics that I'd neglected to achieve as well-rounded an education as I should have. Dr. Horiagon had nurtured a zest for learning in various fields, and our discussions over the years had veered into other topics. The Pareto principle was put forward by an Italian economist who identified a 20 percent input was typically followed by 80 percent output. For example, 20 percent of Pareto's pea plants produced 80 percent of the peas.

"In the case of pharmaceutical design and economics, I would suggest a 10 percent input to 90 percent return on R&D," said Dr. Horiagon. "In other words, the vast majority of the capital to support gene therapy development is coming from 10 percent of the wealthiest individuals."

I was aware that I supported a utopian view of medical treatment, but it was an impossibility for me to break from those who experienced desperate unmet medical need. I did acknowledge, however, that insufficient infrastructure could enhance the suffering in many cases. Gene therapy of type 1 spinal muscular atrophy (SMA) is nothing short of miraculous when given as early in infancy as possible. When given to a child suffering from advanced type 1 SMA, a child in an impoverished country could experience advanced need for medical equipment not available in their country of birth. Ultimately, this could lead to even more suffering in the life of an innocent child and their family.

We talked for hours until the sun set low in the sky and the lobby of our Hilton Garden Inn warmed with the bright banners hanging from the ceiling. We could have continued for several hours longer, but we knew it was time for goodbyes.

"I'll go this way, and you go that way," Dr. Horiagon said.

It was clear we needed to move quickly like one ripping off a Band-Aid. We turned and headed toward the elevator. As I turned back, the doctor was already gone.

A month later, we were traveling in Seattle. Martin was attending a summer camp at DigiPen Institute of Technology. He was knee-deep in the college search process, and we decided to check out his most distant choice first.

Jeremy and I took in the beauty of Puget Sound on a ferry to Bainbridge Island. I was enchanted with the artists' colony and the quaint restaurants and shops. The western hemlock and the western white pine trees fascinated me. I felt I would be inspired to write many novels if left idle in this gorgeous setting.

During our visit, Jeremy was drawn to a rocky beach below a bluff. He wanted to take me with him to explore the shoreline. But the descent presented many obstacles, so I chose to wait on the bluff. As I watched from above, Jeremy explored the rocky shoreline.

Momentarily, my cell phone chirped with an email from Dr. Horiagon. He referenced a link regarding gene therapy. "You should definitely take this into consideration," read his text. In an August 11 article, Novartis had reported that two young children, one from Russia and one from Kazakhstan, had died after receiving gene therapy for type 1 SMA.[2]

I couldn't have chosen a more awe-inspiring setting to ponder the difficult questions regarding gene therapy. From the bluff overlooking the shoreline, Jeremy was in my line of sight—the most important person affected by this most important decision of my lifetime. My custom-made underground gene therapy had idled in my freezer for five months. I had never intended for this shipment to be anything but a test run, a chance to see if the impossible could be achieved. The dosage was not sufficient because I'd ordered a lower dose sample to save money. (10^9 vs. 10^{13}). Upon reading about my plans to potentially breach the ivory tower of medicine, friends in my Gotham Writers Workshop had raised two thousand dollars in GoFundMe donations to allow me to order gene therapy at the concentration of 10^{13}.

Had I not been married, I would likely have ordered the accurate dose of gene therapy and taken it by IV infusion soon after it arrived. I no longer had a fear of death, but this did not imply that I wanted to die. My wish was the opposite, but without a life partner, and a child, I wouldn't need to consider anyone's viewpoint but my own. Faced with very different stakes in this morality equation, I would have chosen and celebrated the chance to help others who currently are experiencing or would experience in the future the genetic hand I had been dealt.

For now, I would consider the timing of placing my next gene therapy order, this time at the clinically relevant dosage. I have no doubt it will occupy a place in my freezer, perhaps wedged between a box of Popsicles and frozen ground beef.

Perhaps, dear reader, I will have an answer for you soon. But, for now, I will leave you to ask yourself, "What would you do if you found yourself in my situation?"

A few weeks later, Dr. Horiagon descended the basement steps of his Highland Parks, Colorado, home, medical file in hand. He wanted to take in the medical reports that sealed his fate on his own terms. Sadly, the MRI scans and physician's notes and dictations defined a very aggressive cancer invading many of the doctor's organ systems.

He had, perhaps, weeks.

I had suspected something was very wrong when I didn't see an email reply once Dr. Horiagon returned from his law school visit in New Hampshire. My suspicions were confirmed when I received an email from his widow, Mai Horiagon, that her husband, Dr. Horiagon, had passed away October 31, 2022.

I had never dreamed the doctor I had turned to when I faced life's medical challenges would face his own tsunami of medical uncertainty and ultimately pass away before me.

12

ATTEMPTING TO SCALE LIFE'S GREATEST CHALLENGE

In the spring of 2024, I made a pilgrimage to the laboratory of Dr. Lori Wallrath at the University of Iowa. I couldn't help but draw a comparison to Michaelangelo's *The Creation of Adam*, as a test tube, filled with wriggling fruit fly larvae, was placed in my hands. For me, the experience was as close as one could get to a "knowing" creator eyeing his handiwork through godlike spectacles. Just inches from my eyes, these simple life-forms—the kind mindlessly swatted as picnic crashers or reduced to a sizzle in a bug zapper—reproduced the symptoms of a muscle disease that had plagued my family for generations. It was hoped that the experiments in the Wallrath lab would lead to greater understanding of my family's genetic disorder—Emery-Dreifuss muscular dystrophy—and perhaps an effective treatment someday.

The fifty or so fruit fly larvae in the plastic test tube, transgenic for my *LMNA* gene mutation and affectionately called "Jill" flies, demonstrated noticeable differences in movement. It didn't take a PhD to sort the transgenic flies from the others. The healthy fly larvae climbed the sides of the test tube with ease. In contrast, many of the mutant larvae could barely burrow out of their food. They rolled helplessly to the side of the test tube, eyeing the vertical challenge that loomed before them like they were staring down Mount Kilimanjaro.

These fruit fly models of our disease held great significance for my family. My brother, Aaron, and I were beset by a much more aggressive progression of our family's genetic disease. While my dad was able to walk until the age of fifty-seven (an average milestone for

Emery-Dreifuss muscular dystrophy patients), Aaron and I lost the ability to walk by the age of thirty-three and required a mobility scooter full-time. Our two other affected siblings appear to be following a similar progression as our father, as they have very mild symptoms.

In 2017, it was discovered that Aaron and I shared a second, very powerful genetic variant in the gene, *SMAD7*. Genetic variants, such as this one, are referred to as modifiers because they alter the effects of a known disease-causing genetic mutation. In this case, the genetic variant in *SMAD7* is not predicted to cause muscle disease on its own, but when paired with a genetic mutation in the *LMNA* gene, the effect on muscle weakness is profound. A modifier may make another genetic mutation more severe or less severe depending on the situation.

Sadly, my brother Aaron passed away in 2019 at the age of forty. We were bestowed with donations to honor his memory. The creation of a fly line possessing the *SMAD7* variant represented a perfect legacy to honor my brother's devotion to academics. Aaron earned a PhD in philosophy and taught university courses for many years. Utilizing the funds, Dr. Wallrath's lab created a custom model of *Drosophila melanogaster* (the scientific name for the fruit fly) that harbored the *SMAD7* DNA sequence variant, which was combined with our *LMNA* mutation to generate what is called the "Jill/Aaron" fly.

This *SMAD7* genetic anomaly, when combined with the *LMNA* variant, magnified the severity of locomotion defects in the flies by 150 percent. While the muscle defects are shared between fruit flies and humans, fruit flies are invertebrates and lack backbones. Therefore, the severe orthopedic deformities that Aaron and I had experienced, particularly with the onset of adolescence cannot be studied on this model organism. I developed severe scoliosis requiring surgical correction by the age of fifteen. Aaron developed a startling contracture of the neck, a deformity that prevented his ability to walk without placement of his interlaced fingers braced behind his neck to bolster his posture forward and provide sufficient observation of the floor.

"I entered junior high as a regular kid but left as an old man," Aaron once remarked to my mother to explain his stark condition.

Aaron managed his very difficult physical affliction from the age of fifteen to twenty-five. Fortunately, after years of searching, we were able to find a surgeon who provided Aaron with a new lease on life following a successful surgery to correct his severe contracture of the neck. Aaron went on to study in Germany for a year. We were overjoyed when Aaron called us to say that while traveling in Italy, he had walked from the Colosseum in Rome to the Vatican with the aid of a walking stick.

Our family's greatest wish has been to spare another family from the trauma our family has endured. Would it be possible to replicate what had gone so wrong in our physical development and test out a treatment or, even better, develop a preventive strategy to ameliorate these tragic symptoms before they surface?

For Dr. Wallrath, the answers couldn't come quickly enough. She devised a genetic experiment to upregulate the amount of *SMAD7* produced by the "Jill" mutant flies. To her amazement, when she observed the larvae following treatment, they were scaling the sides of the test tube as efficiently as the wild-type flies. These same fruit fly larvae that would typically languish in their food had been transformed into visually normal specimens! For Dr. Wallrath, this was a home run she had not been expecting.

When I saw the fly larvae climbing up the side of the tube, I said, "Oh, my gosh, they've never done this before!"

It was awe-inspiring to think that the flies replicating our disease had, with experiments, bypassed the muscular dystrophy that family members and I possessed.

Instinctively, the wheels in my brain moved into translation mode. "What about upregulating the expression in humans? Can you overcompensate and just see what happens—just increase the *SMAD7* and observe what happens?" I asked.

"The fruit fly experiments suggest that this strategy might be successful," said Dr. Wallrath.

It felt triumphant to ask questions of Dr. Wallrath as if I've mastered a foreign language and can speak to her freely as a nonnative speaker. I don't possess the knowledge of a graduate student or a medical student in her lab, but she entertained my question sincerely, for which I was grateful.

"In the fly, it totally compensated," said Dr. Wallrath. "The work that graduate student Nathan Mohar has done shows that *SMAD7* has benefits in muscle. If we over-express *SMAD7* in the fly larvae—remember these are the ones that don't crawl at all, they just barely crawl to the side of the vial—their mobility becomes normal. Thus, there is hope for restoration of muscle function."

In an endorphin-fueled state of optimism, I imagined the affected members of my family being offered the chance to try out this novel therapy. As if she could read my thoughts, Dr. Wallrath described a cautionary tale of this infatuation with gene therapy.

The field of gene therapy had all but been extinguished in 1999 when the field experienced the tragic loss of an eighteen-year-old patient named Jesse Gelsinger in a gene therapy trial. Gelsinger's body experienced what physicians describe as a "cytokine storm," in which the immune system creates a firestorm of attack and the body's organs shut down, one by one.

Over the next fifteen years, scientists studied what had gone so wrong in the case of Gelsinger's death and explored how they could make the field of gene therapy safer. Two important changes were incorporated: First, researchers found that adeno-associated viruses, AAVs, which are smaller versions of the original adenovirus, were better adapted to provide stealth delivery of genetic material than their larger cousins. Second, scientists could select from many versions of AAVs, each fine-tuned to a particular tissue type or types. Fortunately, these new Trojan horse–style tools of gene delivery maintained a preferable safety profile. However, despite these precautions, patient deaths, though a rarity, occurred once again.

"I was surprised that researchers didn't push their failures under the rug," explained Dr. Wallrath. "They [the attendees of the 2024 Muscular Dystrophy Association Conference] went through discussion about patients given gene therapy, and they talked about why those patients died."

The picture she painted was quite like the cytokine storm experienced by some patients exposed to COVID. Despite best efforts, the outer covering of the virus still requires fine-tuning. Scientists are making great gains, such as striving to get gene therapy to a specific tissue. Gene therapy for vision disorders is particularly advantageous because the vision system is much less accessible to the immune system.

"The kidney and liver are the clearinghouses for drugs in the body," explained Dr. Wallrath. "So what they're trying to do is make the AAV vectors containing gene therapy go selectively into the muscle.

Researchers continue to better develop AAV gene therapy to improve targeting in muscle."

In terms of patient deaths attributed to gene therapy, Dr. Wallrath explained that more nuanced issues may be at play—even a minor cold may play a role in an immune-induced attack. As medical reviewers pored over the data, it became clear that when children had some type of infection before receiving gene therapy, the gene therapy became more dangerous and was less successful.

"We've come a long way. We've come further than I anticipated," explained Dr. Wallrath. "With Duchenne [Duchenne muscular dystrophy] and SMA [spinal muscle atrophy] the clinical phase of gene therapy happened so fast. That's kudos to people like yourself and other parents pushing the system because I think that's what helps. This helps to get the attention from the FDA."

Dr. Wallrath praised the Muscular Dystrophy Association (MDA) for finding merit in the startling results uncovered in a single family (the extraordinary effect of a *SMAD7* genetic variant as a modifier in Emery-Dreifuss muscular dystrophy). She, and her lab, have received a three-year grant from MDA to carry out her groundbreaking research.

"MDA sees the value in single family studies. You can learn a lot from single family studies," Dr. Wallrath emphasized.

Dr. Wallrath and her graduate student Nathan Mohar pointed out that not only is the amount of *SMAD7* expression crucial to experiment design but also the timing of expression.

Mohar remains optimistic but addresses important points to consider: "We're doing this in a very clean system genetically. While we are turning on expression of the mutant *LMNA*, we are turning on *SMAD7* overexpression, and they are intimately linked together. That provides us the ability to rescue the crawling defect. So one of the primary issues concerning the treatment of individuals is when developmentally, do you have to overexpress *SMAD7* to have the same therapeutic benefit that you're getting in your very clean laboratory system?"

I could feel the bias in me working up to a fever pitch. I'd seen enough to quell my concerns in these very early experiments. The logical side of me understood that arriving at a place of scientific confidence is a very long process and my emotional perspective of nearly five decades with a genetic disease clouds my judgment. It's this innate desire of research subjects to hit the gas that makes interactions between them and researchers a nonstarter for many scientists and physicians in the field. On the other hand, Mohar raised an additional point as to how my connection with the Dr. Wallrath's lab is atypical.

"It's always tough when you get the news that you have this genetic abnormality and that there's no control over it. Most people don't want to chase it down to the levels that you have," suggested Mohar.

I wouldn't have it any other way.

NOTES

INTRODUCTION

1. The Canadian Press, "Hurdler Lopes-Schliep Sets Fastest Time of Year," CBC Sports, published May 8, 2009, accessed May 20, 2024, https://www.cbc.ca/sports/hurdler-lopes-schliep-sets-fastest-time-of-year-1.778313.

2. Kim Bechert et al., "Effects of Expressing Lamin A Mutant Protein Causing Emery-Dreifuss Muscular Dystrophy and Familial Partial Lipodystrophy in Hela Cells," *Experimental Cell Research* 286, no. 1 (2003, May 15): 75–86, https://doi.org/10.1016/S0014-4827(03)00104-6.

3. Donovan Vincent, "The Amazing Story of Priscilla Lopes-Schliep and the Iowa Mom," *Toronto Star*, January 28, 2016, https://www.thestar.com/news/world/the-amazing-story-of-priscilla-lopes-schliep-and-the-iowa-mom/article_7c704867-09f0-5c39-84e6-377b10334e69.html.

CHAPTER 1

1. Alan E. H. Emery and F. E. Dreifuss, "Unusual Type of Benign X-Linked Muscular Dystrophy," *Journal of Neurology, Neurosurgery & Psychiatry* 29, no. 4 (1966): 338–42, https://doi.org/10.1136/jnnp.29.4.338.

2. Alan E. H. Emery, *My Life* (Devon Oxford: personal memoir, 2011), 41–45.

3. F. E. Dreifuss and Gwendolyn R. Hogan, "Survival in X-Chromosomal Muscular Dystrophy," *Neurology* 11, no. 8 (1961): 734–37, https://doi.org/10.1212/WNL.11.8.734.

CHAPTER 2

1. Sam Roberts, "Lennart Nilsson, Photographer Who Unveiled the Invisible, Dies at 94," *New York Times*, February 1, 2017, https://www.nytimes.com/2017/02/01/world/europe/lennart-nilsson-photographer-embryo-life-magazine-dies.html.
2. Mark Zhang, "A Child Is Born (1965), by Lennart Nilsson," Embryo Project Encyclopedia, published September 17, 2013, accessed May 20, 2024, https://embryo.asu.edu/pages/child-born-1965-lennart-nilsson.
3. Jon Wiener, "The End of the Jerry Lewis Telethon—It's About Time," *The Nation* September 2, 2011, accessed May 20, 2024, https://www.thenation.com/article/archive/end-jerry-lewis-telethon-its-about-time/.
4. Jacquelyn Cafasso, "Recognizing the Signs and Symptoms of Duchenne Muscular Dystrophy," Healthline, published April 27, 2023, accessed May 20, 2024, https://www.healthline.com/health/signs-duchenne-muscular-dystrophy.
5. Santa Saha et al., "Serum Creatine Kinase and Other Profile of Duchenne Muscular Dystrophy and Becker Muscular Dystrophy," *CHRISMED Journal of Health and Research* 8, no. 3 (2021, July–September): 175–81.

CHAPTER 4

1. Gisele Bonne et al., "Mutations in the Gene Encoding Lamin A/C Cause Autosomal Dominant Emery-Dreifuss Muscular Dystrophy," *Nature Genetics* 21 (1999 March): 285–88, https://doi.org/10.1038/6799.

CHAPTER 5

1. Tom Alex, Matt Kelley, and Tom Suk, "Slain Woman Left Colorado to Be Closer to Family," *Des Moines Register*, August 25, 1993, 1A, 10A.
2. Paul M. Renfro, "Fear in the Heartland: How the Case of Kidnapped Paperboys Accelerated the 'Stranger Danger' Panic of the 1980s," Slate, published August 9, 2021, accessed May 21, 2024, https://slate.com/news-and-politics/2021/08/johnny-gosch-eugene-martin-stranger-danger-panic-milk-cartons.html.

3. Bob Berg and Steve Webber, "Kathy Allen (Abduction/Murder): 1987," *Ottumwa Courier*, April 13, 1987, accessed May 21, 2014, https://iagenweb.org/boards/vanburen/documents/index.cgi?read=870648.

CHAPTER 6

1. M. P. Merchut, D. Zdonczyk, and M. Gujrati, "Cardiac Transplantation in Female Emery-Dreifuss Muscular Dystrophy," *Journal of Neurology* 237, no. 5 (1990 August): 316–19, https://doi.org/10.1007/BF00314751.

CHAPTER 8

1. James MacDonald, "The Genetics of First Cousin Marriage," *JSTOR Daily* (August 20, 2018,) accessed May 9, 2024, https://daily.jstor.org/the-genetics-of-cousin-marriage/ (9 May 2024.)
2. Lori Herbert, "When Do Babies Get Their First Eyelashes: A Timeline," *Focus on Your Child* (blog) August 22, 2024, https://www.focusonyourchild.com/when-do-babies-get-their-first-eyelashes/.
3. Kate Marple, "Fetal Development Week by Week," Baby Center, published February 21, 2023, accessed May 9, 2024, https://www.babycenter.com/pregnancy/your-baby/fetal-development-week-by-week_10406730.
4. Wikipedia, "Preimplantation Genetic Diagnosis," accessed May 9, 2024, https://en.wikipedia.org/wiki/Preimplantation_genetic_diagnosis.
5. Gisele Bonne et al., "Mutations in the Gene Encoding Lamin A/C Cause Autosomal Dominant Emery-Dreifuss Muscular Dystrophy," *Nature Genetics* 21 (1999 March): 285–88.

CHAPTER 9

1. Lois M. Freeman, *Betty Crocker's Parties for Children* (New York: Golden Press, 1964), 142–43.
2. Joni Eareckson Tada, *Joni: An Unforgettable Story* (New York: Bantam, 1978).
3. Sudip Saha, "Troponin Testing: What Do Elevated Levels Mean?" Permanente Medicine, published December 16, 2022, https://mydoctor.kaiserpermanente.org/mas/news/troponin-testing-what-do-elevated-levels-mean-2122042.

4. Joseph W. Maricelli et al., "Systemic SMAD7 Gene Therapy Increases Striated Muscle Mass and Increases Exercise Capacity in a Dose-Dependent Manner," *Human Gene Therapy* 29, no. 3 (2018): 390–99, https://doi.org/10.1089/hum.2017.158.

5. Angelo Fichera, "Post Misleads on J&J COVID Vaccine, DNA," Associated Press, February 2, 2023, https://apnews.com/article/fact-check-johnson-johnson-vaccine-adenovirus-dna-101266877973.

6. David Epstein, *The Sports Gene* (New York: Penguin Books, 2013), 106.

7. Jonah Engel Bromwich, "Death of a Biohacker," *New York Times*, May 19, 2018.

8. *Unnatural Selection*, created by Leeore Kaufman and Joe Egender, produced by Lauren Haber and Marc Zahakos, released October 18, 2019, on Netflix.

9. Kirsten V. Brown, "Home-Made Covid Vaccine Appeared to Work, But Questions Remained," *Bloomberg*, October 10, 2023, https://www.bloomberg.com/news/articles/2020-10-10/home-made-covid-vaccine-appeared-to-work-but-questions-remained.

10. Lawrence K. Altman, *Who Goes First?: The Story of Self-Experimentation in Medicine* (New York: Random House, 1987), 1–5.

11. Elizabeth Svoboda, "The Worms Crawl In," *New York Times*, July 1, 2008.

12. Lauren Dillon, "John Hunter: Syphilis, Hubris, and the Great Misbegotten Experiment," Historic Mysteries, published December 9, 2022, accessed May 7, 2024, https://www.historicmysteries.com/science/john-hunter-syphilis/29227/.

13. Altman, *Who Goes First?*, 128, 358–59.

14. Haroon Janjua, "The Horror of Honor Killings," Asia Democracy Chronicles, published March 15, 2022, https://adnchronicles.org/2022/03/15/the-horror-of-honor-killings/.

CHAPTER 11

1. Jill E. Dopf and Thomas M. Horiagon, "Deletion Mapping of the *Aequorea victoria* Green Fluorescent Protein," *Gene* 173, no. 1 (1996): 39–44, https://doi.org/10.1016/0378-1119(95)00692-3.

2. Ned Pagliarulo, "Novartis Reports Deaths of Two Patients Treated with Zolgensma Gene Therapy," BioPharmaDive, published August 11, 2022, accessed July 9, 2024, www.BioPharmaDive.com/news/Novartis-Zolgensma-patient-death-liver-injury/629542/.

BIBLIOGRAPHY

Alex, Tom, Matt Kelley, and Tom Suk. "Slain Woman Left Colorado to Be Closer to Family." *Des Moines Register*, August 25, 1993, 1A, 10A.

Altman, Lawrence K. *Who Goes First?: The Story of Self-Experimentation in Medicine*. New York: Random House, 1987.

Bechert, Kim, Mariana Lagos-Quintana, Jens Harborth, Klaus Weber, and Mary Osborn. "Effects of Expressing Lamin A Mutant Protein Causing Emery-Dreifuss Muscular Dystrophy and Familial Partial Lipodystrophy in Hela Cells." *Experimental Cell Research* 286, no. 1 (2003, May 15): 75–86. https://doi.org/10.1016/S0014-4827(03)00104-6.

Berg, Bob, and Steve Webber. "Kathy Allen (Abduction/Murder): 1987." *Ottumwa Courier*, April 13, 1987, accessed May 21, 2024, https://iagenweb.org/boards/vanburen/documents/index.cgi?read=870648.

Bonne, Gisele, et al. "Mutations in the Gene Encoding Lamin A/C Cause Autosomal Dominant Emery-Dreifuss Muscular Dystrophy." *Nature Genetics* 21 (1999 March): 285–88. https://doi.org/10.1038/6799.

Bromwich, Jonah Engel. "Death of a Biohacker." *New York Times*, May 19, 2018.

Brown, Kirsten V. "Home-Made Covid Vaccine Appeared to Work, But Questions Remained." *Bloomberg*, October 10, 2023, https://www.bloomberg.com/news/articles/2020-10-10/home-made-covid-vaccine-appeared-to-work-but-questions-remained.

Cafasso, Jacqueline. "Recognizing the Signs and Symptoms of Duchenne Muscular Dystrophy." Healthline. Published April 27, 2023. Accessed May 20, 2024. https://www.healthline.com/health/signs-duchenne-muscular-dystrophy.

The Canadian Press, "Hurdler Lopes-Schliep Sets Fastest Time of Year," *CBC Sports*. Published May 8, 2009. Accessed May 20, 2024. https://www.cbc.ca/sports/hurdler-lopes-schliep-sets-fastest-time-of-year-1.778313.

Dillon, Lauren. "John Hunter: Syphilis, Hubris, and the Great Misbegotten Experiment." Historic Mysteries. Published December 9, 2022. Accessed May 7, 2024. https://www.historicmysteries.com/science/john-hunter-syphilis/29227/.

Dopf, Jill, and Thomas M. Horiagon. "Deletion Mapping of the *Aequorea victoria* Green Fluorescent Protein." *Gene* 173, no. 1 (1996): 39–44. https://doi.org/10.1016/0378-1119(95)00692-3.

Dreifuss, F. E., and Gwendolyn R. Hogan. "Survival in X-Chromosomal Muscular Dystrophy." *Neurology* 11, no. 8 (1961): 734–37. https://doi.org/10.1212/WNL.11.8.734.

Emery, Alan E. H. *My Life*. Devon Oxford: personal memoir, 2011.

Emery, Alan E. H., and F. E. Dreifuss. "Unusual Type of Benign X-Linked Muscular Dystrophy." *Journal of Neurology, Neurosurgery & Psychiatry* 29, no. 4 (1966): 338–42. https://doi.org/10.1136/jnnp.29.4.338.

Epstein, David. *The Sports Gene*. New York: Penguin Books, 2013.

Fichera, Angelo. "Post Misleads on J&J COVID Vaccine, DNA." Associated Press, February 2, 2023. https://apnews.com/article/fact-check-johnson-johnson-vaccine-adenovirus-dna-101266877973.

Freeman, Lois M. *Betty Crocker's Parties for Children*. New York: Golden Press, 1964, 142–43.

Herbert, Lori. "When Do Babies Get Their First Eyelashes: A Timeline." *Focus on Your Child* (blog), August 22, 2024. https://www.focusonyourchild.com/when-do-babies-get-their-first-eyelashes/.

Janjua, Haroon. "The Horror of Honor Killings." Asia Democracy Chronicles, published March 15, 2022. https://adnchronicles.org/2022/03/15/the-horror-of-honor-killings/.

MacDonald, James. "The Genetics of Cousin Marriage." *JSTOR Daily*, August 20, 2018. Accessed May 9, 2024. https://daily.jstor.org/the-genetics-of-cousin-marriage/.

Maricelli, Joseph W., Yemeserach M. Bishaw, Bo Wang, Min Du, and Buel Rodgers. "Systemic SMAD7 Gene Therapy Increases Striated Muscle Mass and Increases Exercise Capacity in a Dose-Dependent Manner." *Human Gene Therapy* 29, no. 3 (2018): 390–99. https://doi.org/10.1089/hum.2017.158.

Marple, Kate. "Fetal Development Week by Week." Baby Center. Published February 21, 2023. Accessed May 9, 2024. https://www.babycenter.com/pregnancy/your-baby/fetal-development-week-by-week_10406730.

Merchut, M. P., D. Zdonczyk, and M. Gujrati. "Cardiac Transplantation in Female Emery-Dreifuss Muscular Dystrophy." *Journal of Neurology* 237, no. 5 (1990 August): 316–19. https://doi.org/10.1007/BF00314751.

Pagliarulo, Ned. "Novartis Reports Deaths of Two Patients Treated with Zolgensma Gene Therapy." BioPharmaDive. Published August 11, 2022. Accessed July 9, 2024. www.BioPharmaDive.com/news/Novartis-Zolgensma-patient-death-liverinjury/629542/.

Renfro, Paul M. "Fear in the Heartland: How the Case of Kidnapped Paperboys Accelerated the 'Stranger Danger' Panic of the 1980s." Slate. Published August 9, 2021. Accessed May 21, 2024. https://slate.com/news-and-politics/2021/08/johnny-gosch-eugene-martin-stranger-danger-panic-milk-cartons.html.

Roberts, Sam. "Lennart Nilsson, Photographer Who Unveiled the Invisible, Dies at 94." *New York Times*, February 1, 2017. https://www.nytimes.com/2017/02/01/world/europe/lennart-nilsson-photographer-embryo-life-magazine-dies.html.

Saha, Santa, Anindita Joardar, Sarnava Roy, Tanushree Mondal, Gautam Gangopadhyay, Dibakar Haldar, and Harendra Nath Das. "Serum Creatine Kinase and Other Profile of Duchenne Muscular Dystrophy and Becker Muscular Dystrophy." *CHRISMED Journal of Health and Research* 8, no. 3 (2021, July–September): 175–81.

Saha, Sudip. "Troponin Testing: What Do Elevated Levels Mean?" Permanente Medicine. Published December 16, 2022. https://mydoctor.kaiserpermanente.org/mas/news/troponin-testing-what-do-elevated-levels-mean-2122042.

Schwartz, Ketty, et al. "Mutations in the Gene Encoding Lamin A/C Cause Autosomal Dominant Emery-Dreifuss Muscular Dystrophy." *Nature Genetics* 21 (1999, March 1): 285–88.

Sullivan, Ann M. "*Man and Freedom* Has a New Home." *Mayo Clinic Proceedings* 76, no. 10 (2001 October): 1071.

Svoboda, Elizabeth. "The Worms Crawl In." *New York Times*, July 1, 2008.

Tada, Joni Eareckson. *Joni: An Unforgettable Story*. New York: Bantam, 1978.

Unnatural Selection. Created by Leeore Kaufman and Joe Egender. Produced by Lauren Haber and Marc Zahakos. Released October 18, 2019, on Netflix.

Vincent, Donovan. "The Amazing Story of Priscilla Lopes-Schliep and the Iowa Mom." *Toronto Star*, January 28, 2016. https://www.thestar.com/news/world/the-amazing-story-of-priscilla-lopes-schliep-and-the-iowa-mom/article_7c704867-09f0-5c39-84e6-377b10334e69.html.

Wiener, Jon. "The End of the Jerry Lewis Telethon—It's About Time." *The Nation*, September 2, 2011. Accessed May 20, 2024. https://www.thenation.com/article/archive/end-jerry-lewis-telethon-its-about-time/.

Wikipedia. "Preimplantation Genetic Diagnosis." Accessed May 9, 2024. https://en.wikipedia.org/wiki/Preimplantation_genetic_diagnosis.

Zhang, Mark. "A Child Is Born (1965), by Lennart Nilsson." Embryo Project Encyclopedia. Published September 17, 2013. Accessed May 20, 2024. https://embryo.asu.edu/pages/child-born-1965-lennart-nilsson.

INDEX

AAV gene therapy, 196–97
abduction, 80
Achilles tendons, 97
ADD. *See* attention deficit disorder (ADD)
AIDS, 154
aloneness, 56, 111, 138, 173
Altman, Lawrence, 156
Anderson, Robert, 18, 23, 32–34
Appalachian medicine, 11–14
attention deficit disorder (ADD), 161

Barrymore, Drew, 106
"A Benign Type of X-Linked Muscular Dystrophy with Unusual Features," 13
Betty Crocker's Parties for Children, 146
bioethicists, 189–90
biotech supply company, 163
biotech websites, 153

bizarre theory, 5
"Blue Star Homes," 80
Brown, Emmet, 152

cancer, 187; gene therapy for, 183
Care Bear Collection, 54
Care Clinic, 89
Carrion, Daniel, 156
A Child Is Born, 17
"citizen scientist," 1–2
Clark, Petula: "Downtown," 113
claustrophobic, 142
Communion, 27
COVID, 149, 151–53, 155, 181, 182, 187, 188, 196
Crawford, Cindy, 7
creatine kinase (CK), 33, 35, 162
CRISPR technology, 154
custodial care, 148
"cytokine storm," 196

dating: agency, 106, 107, 109;
    problem with physical disability,
    106–7, 109, 122
DigiPen Institute of Technology, 191
disability, 27
disease-causing genetic mutation, 194
"Downtown" (Clark), 113
Drake University, 28–29, 83, 84,
    183
"The Drama of Life Before Birth,"
    17
Dreifuss, F. E., 13
Duchenne muscular dystrophy, 34
dysphoric, 64

Emery, Alan, 11–14
Emery-Dreifuss muscular dystrophy
    (EDMD), 3, 4, 6, 14,
    57, 74, 89, 91, 92, 136,
    139; diagnosis of, 172;
    medications, 139–40
Epstein, David, 5, 160; *The Sports
    Gene*, 153–54

follicle-stimulating hormone, 135
Forensic Files, 77, 78
frozen joint, 54, 60

Gelsinger, Jesse, 196
gene therapy, 153, 162–63, 188, 190,
    196; for cancer, 183; patient
    deaths attributed to, 197;
    of type 1 spinal muscular
    atrophy, 190; for vision
    disorders, 196
genetic disease, 182, 193
genetic disorder, 174; osteogenesis
    imperfecta, 109
genetic medicine, 13
genetic risk, 127, 128

genetic testing, 71, 74, 76, 153,
    166–67; muscular dystrophy
    without, 159
genetic therapy, 152, 162
genetic variants, 74, 97, 194
GoFundMe donations, 192
gonorrhea, 157
Gordon, Hymie, 34–37
Gosch, Noreen, 132
Gotham Writers Workshop, 192
Gower's sign, 33–35, 150
Greater Columbus Convention
    Center, 127
Great Iowa Treasure Hunt, 75
green fluorescent protein (GFP), 183,
    184
growth spurts, 61

hedge fund, 190
hereditary disorders, 3, 4, 36
hereditary muscle disease, 150
Hogan, Gwendolyn R., 13
honor killings, 160
Horiagon, Tom, 181, 183–92
Human Gene Therapy Research
    Institute, 183
human suffering, cruel depiction of,
    53
Hunter, John, 157

immune-induced attack, 197
Invitae, 159
"Iowans for Life" sign, 134
Iowa State Fair, 17, 75, 78
Iowa State University, 116

Jackson, Samuel L., 109
Jerry Lewis Labor Day Telethon, 120
"Jill" flies, 193–95
Joel, Billy: "Uptown Girl," 113

Johns Hopkins, 12, 14, 181
Jones, Lolo, 4
Joni: An Unforgettable Story, 147–48
"juvenile, proximal muscular dystroph," 64

Keats, Mardella, 48
ketamine, 154
Khan, Iman, 178
Kirmani, Salman, 166
Klaussen (Dr.), 63–65

Lange, Patricia, 78
Laura (Dr.), 137
law career, 82
"Leap Year Twins," 43
Lewis, Jerry, 31
Life (magazine), 17
limb-girdle muscular dystrophy (LGMD), 164–67, 176, 177
LMNA gene mutation, 193, 194, 197
loneliness, 111
Lopes-Schliep, Priscilla, 1–9, 149, 155
lordosis, 12

malnutrition, 12
Mayo Clinic, 89, 118; phlebotomy department, 120
McConaughey, Matthew, 154
McKusick, Victor, 12, 13
MDA. *See* Muscular Dystrophy Association (MDA)
medical memoir, stakes of, 175
mediocrity, 141
Meeks, Miki, 46
Mendelian Inheritance in Man, 12
Mohar, Nathan, 195, 197, 198
mother–daughter relationship, 94
muscle weakness, 177

muscular dystrophy, 14, 31–35, 37, 71, 127, 148; form of, 167; without genetic testing, 159
Muscular Dystrophy Association (MDA) Conference, 31, 89, 118, 196, 197
muscular dystrophy forum, 149
Mychasiw, Kris, 5
myopathic muscle biopsy, 73
Myszewski, Mike, 88, 89
The Naked Don't Fear the Water, 174

National Institute of Mental Health, 141
National Institutes of Health, 13
National Society of Genetic Counselors annual conference (2017), 127
neuromuscular disease, 33
Nilsson, Lennart, 17
The ODIN, 154
"oldest daughter syndrome," 182

Online Mendelian Inheritance in Man (OMIM), 12
Oroya fever, 156
orphan disease, 185, 186
OSHA violations, 71
osteogenesis imperfecta, 109
Ouija board, 58–59

Pakistan, traveling into, 169–79
Pareto principle, 190
partial lipodystrophy, 7
paternalism, 185
Peruvian Medical Society, 156
photography, 66–69
physical disabilities, 161; dating problem with, 106–7, 109, 122

physical distress, 25
physically disabled person, life of, 145–46
physical therapists, 54
physical therapy, 162; exercises, 55; sessions, 45
Piper, Donald, 78
polio vaccine, 157
Pritchard, David, 156–57
pro-life advertising campaign, 133

rape, 78, 82
robbery, 84–85
Roe v. Wade, 17

Salk, Jonas, 157
SARS-CoV-2 virus, 155
scoliosis, 62, 67
The Secret of NIMH, 141
sexually transmitted diseases (STDs), 157
Shriners clinic, 98
SMAD7 genetic variant, 163, 194, 195, 197
social networking project, 58
Social Security Disability income, 161
spinal muscle atrophy (SMA), 162
The Sports Gene (Epstein), 153–54
Spring 1972, 16–27
state-run Medicaid programs, 190
Stearman, Kathy, 169, 171

summer (1992), 83–90
Sweeney, H. Lee, 153
syphilis, 157

Tada, Joni Eareckson, 147–48
Toniolo, Daniela, 71
Traywick, Aaron, 154
type 1 spinal muscular atrophy (SMA), 190

United States: legalized abortion in, 17; weak muscles, 164
Unnatural Selection, 154
un-walking, 91–103
"Uptown Girl" (Joel), 113

vision disorders, gene therapy for, 196

Wallrath, Lori, 144, 193–98
Ward, Clarrisa, 175
White, Sue, 43, 45–47
Who Goes First? (1998), 156
Winter (1979), 29–41
Woolery, Chuck, 105
World War II, 12

Zayner, Josiah, 154, 155
Zolgensma, 162
Zoom Gotham Writers Workshop, 163

ABOUT THE AUTHOR

Jill Viles holds a master's in creative writing from Iowa State University and is an annual attendee of the University of Iowa Summer Writing Festival. Her writing has appeared in *Johns Hopkins Magazine*, and her scientific work has been published in the journals *Gene* and *Cells*. In the course of researching her book on her family's rare muscle disorder, she has

collaborated with scientists at the University of Iowa performing gene therapy on mutant fruit flies.

Jill lives in rural Iowa with her husband, Jeremy, and a lively golden retriever, Yoshi. Recent empty nesters, they are enjoying visits to see their son, Martin, at the University of Iowa. Jill lives by the adage that there's nothing a Coke with grenadine flavoring can't remedy, and she's charmed by both the crackle of a warm fire and the crack of a newly opened book.